In-Game

The world will be Tlön.

—Jorge Luis Borges

Contents

Acknowledgments

The greatest challenge of this project has been the multidisciplinarity required to address player experience in a holistic manner. Even with a working knowledge of the fields in question, one can scarcely be an expert in them all. I have been most fortunate to have had input and feedback from people who come from a variety of disciplines and backgrounds and who have helped to make this book possible.

First of all, I am indebted to the mentors who have nourished my thoughts, guiding and supporting me through the PhD process: to Warwick Tie and Joe Grixti for their stimulating conversations and their belief that this was a project worth undertaking; to Brian Opie for endless hours of discussion, argument, and encouragement; and, above all, to Ivan Callus, for his life-long guidance, faith, and inspiration.

Second, I am deeply grateful to all those colleagues who have been patient enough to discuss and review the material that would end up becoming this book: Thomas Malaby, Timothy Burke, Espen Aarseth, T. L. Taylor, Rune Klevjer, Stephan Günzel, Bjarke Liboriussen, Malcolm Zammit, and finally Pippin Barr, who proved to be an indispensable source of critical feedback and an outstanding editor. I would also like to thank Chris Borg Cutajar for his work on the Web site and Steffi Degiorgio, who designed the illustrations and the book cover.

Above all, my gratitude goes to Anne Hamarsnes, whose patience and love have kept me sane throughout the process and whose calm wisdom has been a guiding light on its darker days. I dedicate this book to her.

Introduction

Although digital games have been around for just over three decades, their presence in popular culture has become pervasive. Since the arrival of the first commercially available consoles like the Atari VCS and computers like the Spectrum ZX, Amstrad CPC, and Commodore 64, digital games have rapidly altered the landscape of media entertainment.

In my home village of Gzira, Malta, as soon as school was over, the roofs and streets used to be covered with swarms of kids playing football. As soon as the Commodore Amiga hit the shops, the streets and roofs emptied. Now, when the school bell rang, the swarms of kids had joysticks strapped to their bags instead of footballs. The football tournaments moved indoors to the flats of the lucky few who owned an Amiga. In the limited space of a densely populated town like Gzira, digital gaming provided sites for play that were not punctuated by the interruptions of cars driving through our chalked-in football pitches. The migration indoors caused by the Amiga was a felicitous moment not only for us, but for our worried parents and all the car drivers who were fed up with honking their horns and waiting for us to clear the street.

Aside from enabling us to transcend the practical limitations of our environment, digital games[1] became popular because they transported our imaginations to the places represented on screen. We no longer had to imagine landscapes of forests and mountains to roam in; they were right there in front of us. Since such terrain features were lacking on the island, we associated them with the fantasy literature we were so fond of. For Maltese kids who have never been abroad, forests and mountain ranges are the stuff of fantasy itself, and having the chance to inhabit those landscapes, albeit in an abstractly represented simulation, was an alluring part of digital games. On the other hand, digital games also offered a structure for competitive play in games such as *Kick-Off* (Dini, 1989), in which the rules of the game, upheld by the machine, offered a more level playing field. Both broad forms of engagement often occurred in shared settings, with single-player games involving one player[2] navigating the environment and the others making suggestions or ultimately wrestling the joystick out of the player's hands.

These early encounters with computers made my friends and me feel comfortable in the cybernetic circuits that would become part of our everyday adult lives. Games introduced us to a symbiotic relationship with machines that we took for granted. We grew up acclimatized to a technology that would not only entertain us and facilitate our work life, but would ultimately change the way we thought and operated socially. I was part of the first generation of children to dive with great zeal into the alluring cybernetic circuit that games provide. Of course, digital games were not the only sites for engaging encounters with computing machines, but they were definitely, even in the late 1980s, the most widespread form of *involving* cybernetic engagements. Such a claim invites a question: What exactly makes digital games so involving? This book suggests an answer.

This relationship between player and game yields a novel form of engagement that calls for a dedicated theoretical understanding. This is not to say that perspectives on involvement developed in the context of other media are irrelevant, but that the specific characteristics of digital games, or at least a subset of digital games, need a more thorough analysis than such theories can provide. The study of player involvement in digital games requires an expansion and, at times, a rethinking of such theories. This book offers a perspective on player involvement in digital games that takes into account the games' specific qualities and characteristics. This perspective is embodied in the player involvement model that addresses digital game involvement ranging from general motivations and attractions to the detailed analysis of moment-to-moment involvement in gameplay.

An important component of player involvement is the shortening of the subjective distance between player and game environment, often yielding a sensation of inhabiting the space represented on-screen. This phenomenon is generally referred to in terms of *presence* and *immersion*. The latter is the more commonly used term in popular and academic discussions of game involvement, but its general use has diminished in analytical value. The term *presence* is similarly affected, with the main writers in the field of presence theory often using the term with divergent or even conflicting meanings. This book will therefore examine the application of these two terms and propose a more precise conceptualization of the phenomenon that is specific to game environments. Rather than conceiving of presence or immersion as a stand-alone experiential phenomenon, this book sees it as a blending of a variety of experiential phenomena afforded by involving gameplay. These experiential phenomena are individually described by the player involvement model and in their more intensified and internalized blends have the potential to culminate in *incorporation*, this book's answer to the presence/immersion quandary. Although incorporation is ultimately a metaphor like presence or immersion, it avoids a number of problematic connotations that are present in the other two terms and, more importantly, provides a more robust concept for researchers to build on.

A Brief Overview

Digital games constitute a broad family of media objects, some of which diverge so much in their constituent characteristics that they cannot all be taken as one homogeneous mass. Although we attach the label *game* to both *Grand Theft Auto IV* (Rockstar North, 2008) and *Tetris* (Pajitnov, 1985), the differences between them are so significant that any discussion that considers them as equivalent media objects is prone to make generalizations that impede analytical rigor. Chapter 1 discusses the difficulties that researchers in game studies have experienced in defining games, and describes their constituent elements without restricting games to an essentialist definition. This framework will form the foundation of how digital games will be viewed in the rest of the book. It explains the rationale for the structure of the model and discusses the theoretical foundations and assumptions that underlie its formation.

The second chapter considers the related phenomena of presence and immersion and the way these have been used in presence theory and game studies, respectively. Vagueness and confusion have surrounded these terms, and this chapter outlines four challenges for understanding the experiential phenomenon they refer to in the context of game research.

The third chapter introduces the player involvement model developed through qualitative research. It explains the relationship between its two constituent temporal phases: macro, representing off-line involvement with the relevant game; and micro, representing moment-to-moment involvement during gameplay. After giving a brief outline of attention, the chapter addresses the difference between game involvement and involvement with other media such as film and literature. The chapter also divorces the issue of player involvement from the question of fun and the concept of the magic circle.

Chapters 4 through 9 describe in detail the six dimensions that constitute the player involvement model. Each of these is considered on two temporal phases of engagement, macro- and micro-involvement. Macro-involvement encompasses all forms of involvement with the game when one is not actually playing. These include the initial attraction to the game, reasons for returning to it, participation in the community it fosters, and other off-line plans and thinking that surround the actual instance of gameplay. The micro-involvement phase describes the qualities of moment-to-moment involvement within the respective dimensions. It deals with six dimensions of player involvement: control and movement (*kinesthetic involvement*), the exploration and learning of the game's spatial domain (*spatial involvement*), co-presence, collaboration, and competition with other agents (human or AI) that inhabit it (*shared involvement*), the formation of an ongoing story and interaction with the scripted narrative written into the game (*narrative involvement*), the affect generated during gameplay (*affective involvement*), and the decision making undertaken in the pursuit of both game- and self-assigned goals (*ludic involvement*).

Chapter 10 discusses the phenomenon of incorporation. This chapter will argue that incorporation's specific formulation avoids the four conceptual challenges outlined in chapter 2. It also argues for incorporation as a more accurate metaphor than presence and immersion. The chapter ends with an examination of how incorporation emerges from the combination of several dimensions of the player involvement model through two examples of incorporating experiences described by research participants.

Overall, this book provides a new representation of involvement in digital games and, in the process, builds an argument for rethinking the concept of immersion as a multifaceted experiential phenomenon.

Critical to a precise reconceptualization of the phenomenon is my intro-duction of a new term, *incorporation*, which can account more satisfactorily for the complex range of factors that make up the sense of virtual environ-ment habitation, and can therefore provide a more productive concept for researchers and practitioners in various fields to work with. The player involvement model is intended to provide a common, holistic framework that will facilitate further research in the area.

1 Games beyond Games

Games can be approached analytically from a variety of perspectives. Broadly speaking, however, research tends to emphasize one of three key areas: the formal aspects of games as media objects, experiential and subjective aspects of gameplay, and sociocultural aspects of gaming communities. These are not insulated categories, but rather will necessarily draw on insights from one another while privileging a primary viewpoint, usually as a consequence of the researcher's academic background. This book focuses chiefly on the experiential aspects of games, but the other two perspectives will necessarily inform our exploration of player involvement—it is often the case that the formal and higher-level social issues surrounding games will shed light on the experiential.

Before we can consider the challenges posed by analyzing the player experience of involvement and immersion, we must first address some more fundamental questions about the domain itself. In this chapter I will delineate those members of the broad and varied family of games that will primarily be discussed in this work and provide a description of their constituent elements. Such a description will clarify what is being referred to in the rest of the book when central terms such as *players*, *rules*, and *environmental properties* are used.

The Complexity of Games

Games are a complex social phenomenon that eludes holistic categorization. Attempts to formulate stable, universal definitions of games inevitably fall short of the mark, leaving important aspects of particular games unaccounted for. Yet these omissions can often be as instructive as the ground covered by attempts at definition, reminding us of the multiple perspectives that are relevant to understanding the role of games in social reality. Games reflect aspects of the society and culture that made them while contributing to that society in the process; as a result, understanding them is a recursive process of exploration into collective knowledge and social practices.

To further complicate this process of understanding, the performance of a game occurs in two, often simultaneous, domains: the player's subjective experience, and the visible practice of playing. Gameplay includes actions ranging from moving a piece on a game board, pressing a sequence of buttons on a controller, or sprinting, ball in hand, toward a distant white line. Most importantly, a game becomes a game when it is played; until then it is only a set of rules and game props awaiting human engagement.

The centrality of human subjectivity in the game process lies at the very heart of the challenges game theorists face in the process of their analysis. These difficulties are not aided by the fact that the term *game* includes a wide variety of disparate activities. Although poker, fencing, and *Grand Theft Auto IV* (Rockstar North, 2008) all fall under the general heading of games, each entails a very different form of engagement. Digital games, especially, create an added level of complexity.

Although for ease of reference we call *Grand Theft Auto IV* a game, it may be more accurate to consider it as a virtual environment with a number of games embedded in it and a linear storyline that players can progress through by completing a sequence of gamelike activities. When a player or players enter *Grand Theft Auto IV*'s Liberty City, they can engage in prepackaged games that have been coded into the system or they can decide to create their own games within the virtual playground. The rules of the games they play are thus either upheld by the software or agreed upon socially (as is the case with nondigital games). Further, players may choose to interact in ways that are not gamelike at all, perhaps going for a scenic drive or walk with their friends. In short, not all interactions with the objects we call games result in gamelike activities.

Games as Families

Wittgenstein (1997) suggests viewing games not as a rigidly defined set but as a family whose members share some "family resemblances." The strength of this conception is that it does not require a single list of characteristics to run through all the objects and activities we call games, but instead generates a collective concept based on the overlaps between various members of the family. Although boundaries can be drawn for the sake of analysis, we must be aware that such boundaries are artificial. When we outline the characteristics particular to a set of games, these characteristics need not identify games as a whole, but should identify a subset of the larger family called games.

This analytical specificity is needed within game studies to avoid situations where a disagreement occurs based not on the actual claims being made, but on the exemplars to which those claims are being applied by different parties in the discussion—what Arne Næss (2005, 64) calls a "verbal disagreement." While well-founded and courteously argued disagreements are healthy for the growth of a new field, those based on verbal disagreements fail to make a productive contribution because there is no consensus as to what is being discussed. A close reading of the central debates that have arisen within game studies gives ample indication that such a problem already exists. In the next chapter, we will discuss one such disagreement about the concepts of presence and immersion. When one analyst is building a critique of such terms using *Tetris* (Pajitnov, 1985) as his or her main exemplar, and others are commenting with games like *Half-Life 2* (Valve Software, 2004) in mind, the conversation cannot really move forward because the forms of engagement afforded by the latter are radically different from those afforded by the former.

There have been a number of debates within game studies that have been complicated unnecessarily by a lack of agreement upon the exact subject of discussion. The involved parties might be discussing *games* without making it clear which members of the family of games, virtual environments, or hybrids thereof they are actually considering in their analysis. The rest of this chapter will outline a descriptive framework that identifies the matrix of components that combine to form games and virtual environments. The types of games addressed by the player involvement model outlined in this book are found at the intersection of these two families of objects.

Process versus Object

One important distinction that can be made when discussing games is between game as object and game as process. A board game like *Settlers of*

Catan (Teuber, 1995) is both a set of material objects and rules as well as an activity afforded by those objects and rules. These rules are intended for interpretation and deployment by a group of players in their associated sociocultural context. We can discuss various aspects of the game as object in isolation from the actual situated playing of that game. In relation to *Settlers of Catan*, one can comment on the visual qualities of the hexagonal board pieces or the color scheme used in its deck of cards. One may be critical of the value of the robber in the game, which blocks players from drawing resources from the tile on which it is placed. A genealogy of board games may consider the influence of *Settlers of Catan* on subsequent board game design, and so on.

This division of object and process can also be applied to digital games. The rules are coded into the game instead of being upheld by players, and the material objects involved in its enactment are the software and hardware machines that run them instead of actual game pieces, but the consideration of the game as a tangible object separate from its actualization through interaction with the player remains the same. In the case of digital games, the object is described by the code and the material medium that contains the code. Although considerations of the game as object tend to have important implications on the actual gameplay, we need to acknowledge that they represent only a partial or incomplete view. The dormant code, board pieces, or rule set present a potential that is actualized during gameplay.

This brings us to the second perspective: games as processes. Theorists such as T. L. Taylor (2006) and Thomas Malaby (2007) have recently made strong arguments in favor of a processual approach, possibly as a reaction to a number of prominent texts that focused more on the game as object than as process. Malaby argues: "One of the first things we must recognize is that games are processual. Each game is an ongoing process. As it is played it always contains the potential for generating new practices and new meanings, possibly refiguring the game itself" (8).

The term *processual* refers to the potential for variation in a game's enactment at every engagement and favors a dynamic and recursive view of games. A processual perspective suggests that the identification of persistent features of games is continuous with other domains of experience. This means that games need not be conceptualized as somehow experientially separate, as is implied by the notion of the "lusory attitude" (Suits,

1978, 55) taken up by some theorists in game studies (Salen and Zimmerman, 2003; Juul, 2005) and discussed in chapter 3. Malaby (2007) formulates games as processes that create carefully designed, unpredictable circumstances that have meaningful, culturally shared, yet open-ended, interpretations. Therefore, both the game practice and the meaning it generates are subject to change.

Digital Game Elements

Rather than adhering to an essentialist definition of games, such as those forwarded by theorists like Jesper Juul (2005) or Katie Salen and Eric Zimmerman (2003), this book follows Wittgenstein in viewing games as members of an extended family that share resemblances. We will focus on a subset, or group of subsets, of the game family that occur within virtual environments. Virtual environments and digital games share some common elements that interact to express different configurations of games. These elements are the human player, the representational sign, the structural properties of game and environment, and the material medium which instantiates the combination of these elements. A description of each of these elements follows.

The Player

Here, as in the rest of the book, I will be using the term *player* to refer to the human agent, or agents, that engage with the game system. The use of the term *player* should not, however, be limited to the characteristics commonly associated with play. I am not here subscribing to a notion of play that prescribes a particular experiential disposition, such as *playfulness* (however that is conceptualized), to the human agent engaging with the game. I am using the term *player* instead of *human agent* to conform with the convention within game studies. From the game-as-object perspective, the player is conceived as an ideal, or implied, player. Turning to the game as process, the player is conceived as the actual, active player and the set of practices she deploys in interacting with the game world and game system. These practices are always considered in relation to the social and cultural contexts of the player and have an important formative role in the individual's disposition prior to and during engagement with the game.

In the enactment of the game as process, it is often the case that different players interacting with the same structure, signs, and medium actually experience a different game. I might be playing a multiplayer death-match round in *Call of Duty IV* (Infinity Ward, 2007) following conventional rules—that is, trying to help my team score as many kills as possible while not giving away kills to the enemy. Meanwhile, somebody else on my team might be playing a different game: striving to kill an opponent while jumping off the highest building on the map. Although we are both interacting with the same game object, the resultant game process is different enough to be called a different game. The goals of the suicidal player are not only different but are contradictory to mine.

The Representational Sign

The second element of the framework, the representational sign, refers to the more general sense of a signifying entity, whether this is alphanumeric text, imagery, or sound. The *representational sign* therefore refers to the interpretable, representational elements that players read in order to be able to interact with the game. In the case of digital games, the representational sign might be made of the same code that dictates the behavior of AI agents or the material density of a wooden fence, but for the sake of analysis it makes sense to separate these two configurations of code since they perform very different functions in the game object and process.

Coded Rules

In *Cybertext* (1997), Espen Aarseth argues that in ergodic media the function of the "surface sign" (40) has a strong relation to the mechanical operations of the internal code that generates it. Aarseth distinguishes the forms of processual production of cybernetic signs in ergodic media by referring to the relationship between code and its interpretative surface as "nontrivial" (1): they are mutually intrinsic. In contrast, there are a number of texts that are actualized as the result of more than one level of textuality. A literary text read out loud, for example, is an audible derivative of the printed book. Aarseth calls such examples "trivial" (1), meaning that one level of manifestation is a derivative of the other. The nontrivial relationship between the code and the interpretable sign is what makes the cybernetic sign of a different order from its print predecessors.

Rules are, in one form or another, a common denominator in all members of the game family and thus are always found in games. The shape of rules, however, varies from game to game and is often modified by a player's or a community of players' perspective on the game. In analog games, rules are stipulated by the game, but it is up to the players to follow them, making rules dependent on social convention. This often creates situations in which rules are negotiated by the players and altered to suit their whims.

In the case of digital games, rules are coded into the software and thus are harder to modify. Players would need to amend the actual code of the game in order to change the rules. In multiplayer games, however, we often see a coexistence of coded and socially negotiated rules. A particular clan in *Counter-Strike: Source* (Valve Software, 2004) might use the standard coded rules written into the game but not allow sniper rifles on the servers used to host its matches. The player would still be able to purchase the banned rifles but would receive a warning, and would usually be banned from the server if they did so again. Negotiation still occurs in environments with coded game rules, but often these will still have an element of conventional rules that have been developed by the community.

(Simulated) Environmental Properties

Simulated environmental properties are found whenever a game takes place in a constructed environment that models physical properties. This is not limited to computer-generated environments, as pen and paper ones can also be simulated through game systems. For instance, the fact that the game system of a tabletop RPG indicates that a three-meter fall onto solid ground will yield 3D6 points of damage to the sufferer gives the world a physical structure that grants additional body to the shared mental image, turning it into a simulated model. Computer-generated environments have their physical properties hard-coded into them. Bricks in *Call of Duty IV* (Infinity Ward, 2007) have a certain density that will resist 9mm pistol rounds, but they are penetrated by the larger 5.56mm rounds of an M4A1. Similarly, player avatars can run, walk, and crawl at defined speeds. All these environmental details influence the rules of the game and are crucial in creating a balanced and enjoyable game experience. But again, the difference between a computer-generated environment and an analog one is that in the former the environmental mechanics are upheld by the computer, while in the latter they are maintained by the players in accordance with a prescribed, and often modified, system.

The complexity of simulated environmental properties varies from one virtual environment to another, ranging from simple abstractions of a handful of physical parameters, as in *Pong* (Atari, 1972), to a more complete (although always partial) simulation of aspects of the physical world, as in games like *Crysis* (Crytek Frankfurt, 2007) and *Call of Duty: Modern Warfare 2* (Infinity Ward, 2009).

Material Medium

The specificity of the material instantiation of the game must be taken into consideration. Even if the same game is being discussed, its incarnation on the Playstation 3 rather than on a PC will influence its form and experience to varying degrees. Playing a real-time strategy game using a Playstation 3 controller makes for a very different game than playing the same title on a PC using a mouse, for example. Different types of hardware also support different social contexts for play. Nintendo DS systems, for instance, are handheld devices small enough to fit into a jacket pocket and easily connected via infrared ports, permitting a wider variety of contexts and thus different experiences, than, for example, a home PC enables.

This range is even more marked when we consider the rules of board games expressed in code. Although the rules remain the same whether I am playing *Settlers of Catan* (Teuber, 1995) on the computer or with the board game itself, the practice and experience of playing the game will be different. The lack of a tangible board on a table, resource cards held by players, and game pieces creates a markedly different incarnation of *Settlers*. Whether this can or should be called a different game altogether is less important than having an adequate analytical tool to account for the differences.

Digital Games as Hybrids

The complexity brought about by the advent of digital gaming is continuing to grow. The most significant source of this complexity arises from the fact that a considerable proportion of what are called games nowadays are in fact extended virtual environments which contain a game or multiple games within them. If we had to apply the framework to a game like *Grand Theft Auto IV* (Rockstar North, 2008), for example, we would see a number of different clusters of game rules all simultaneously present in the same overall environment, rather than a single game that would adequately explain them all. The material medium, signs, and environmental properties that form the virtual environment remain the same for all the embedded games, but the conventional (socially agreed upon) rules, coded rules, and player aspects could change in each case. Each game embedded in the environment has its own coded rules and, in the case of multiplayer games, also has its own conventional rules.

In Wittgenstein's (1997) terms, contemporary digital games like *Grand Theft Auto IV* (Rockstar North, 2008), *World of Warcraft* (Blizzard Entertainment, 2004), and *Half-Life 2* (Valve Software, 2004) are members of two families that mingle resemblances. They contain the features of both virtual environments and games, and thus form a subfamily derived from both. It is no surprise that game definitions that have tried to account for both digital games and board games have struggled to cater to both forms. As Ryan (2006) argues, digital games (and tabletop role-playing games before them) have enabled the combination of traditional games' rule systems with the fictional and narrative aspects of the media that preceded them.

To account for the hybrid nature of games in virtual environments, the rest of this book will refer to those members of the game family that are set in virtual environments as *virtual game environments*, or *game environments* for short. Virtual game environments, although a somewhat cumbersome term, accounts for the intersection of the two broad families of virtual environments and games, distinguishing those games that occur within virtual environments from those that do not. Digitized versions of card games like hearts or poker, or puzzle games like crosswords, Sudoku, and the like, are not forms of virtual game environments and will thus not be considered at great length in the rest of the book. Instead, I will focus primarily on games that present the player with a virtual world in which to participate in a variety of activities, as do games such as *Half-Life 2* (Valve Software, 2004) and *World of Warcraft* (Blizzard Entertainment, 2004).

2 Immersion

The prone body connected to a virtual reality machine via implanted neural jacks is a staple image of cyberpunk movies. The figure would seem to be dead were it not for the occasional twitch and spasm betraying the possibility that it is, in fact, dreaming. This is not your average dream, however; it is lucid dreaming on demand, a pay-per-act performance inside a virtual world so compelling it is challenging to distinguish it from reality itself. This popular trope in cyberpunk literature and film has also found traction in public discourse, marketing, and academic discussions surrounding games.

Virtual reality technology brought with it the promise of making such experiences an accessible part of everyday life. It was seen as the culmination of a long history of the creation of virtual environments that attempted to give viewers the sensation that they were actually *there*. According to Oliver Grau (2003), this tendency to create "hermetically closed-off image spaces of illusion" (5) has a rich history that can be traced as far back as 60 BC, when some Roman villas had rooms dedicated to the simulation of another world. The notion of creating an all-encompassing media experience was also a concern of André Bazin, whose influential 1946 essay "The Myth of Total Cinema" argued that the ultimate goal of cinema and all techniques of mechanical reproduction was the creation of "a total and complete representation of reality ... a perfect illusion of the outside world in sound, colour and relief" (Bazin, 1967, 20).

Virtual reality offered the opportunity not only to be visually surrounded by the representational space, but also to move and act within it. Technologies attempting to realize this dream were developed as early as 1965 by Ivan Sutherland, who created a working prototype of what he called "The Ultimate Display," a bulky version of today's head-mounted displays (Sutherland, 1965, 1968). Sutherland's creation gave a new meaning to the notion of "being there" that surpassed even the most compelling representational media. This experiential phenomenon has been referred to most popularly by the terms *presence*[1] (Minsky, 1980) and *immersion* (Murray, 1998).

Although there has been consensus that the experience of presence or immersion is important, there has been confusion over precisely what the terms mean. This disagreement has hampered progress in their conceptualization (Slater, 2003; Waterworth and Waterworth, 2003) and has undermined their utility (Ermi and Mayra, 2005). This is especially problematic for the study of games since, as a number of theorists have argued, the experiential phenomenon referred to by the terms is a crucial part of the player experience (Ermi and Mayra, 2005; King and Krzywinska, 2006; Tamborini and Skalski, 2006; Brown and Cairns, 2004; Jennett et al., 2008). Of the two terms, *immersion* is particularly awkward because it has also been applied to the experience of non-ergodic media such as painting (Grau, 2003), literature (Nell, 1988), and cinema (Bazin, 1967), all of which provide forms of engagement that are qualitatively different from those of game environments.

Presence and Immersion in Presence Theory

Presence is derived from *telepresence*, a term coined by Marvin Minsky (1980) in his paper "Telepresence." In the paper, Minsky describes how operating machinery remotely can lead to a sense of inhabiting the distant space. This sense of presence is created through a combination of the operator's actions and the subsequent video, audio, and haptic feedback. A term was needed to account for the awareness of the potential to act within two spaces: the physically proximal and the physically remote. Minsky's primary concern was the design of hardware that would create a strong enough sense of presence to facilitate teleoperation: "The biggest challenge to developing telepresence is achieving that sense of 'being there.' Can telepresence be a true substitute for the real thing? Will we be able to couple our artificial devices naturally and comfortably to work together with the sensory mechanisms of human organisms?" (Minsky, 1980, 48).

The issues that Minsky raised became important not only in the field of telerobotics, but more generally in virtual reality technology, where fostering a strong sense of telepresence had long been a high priority. These concerns were seen as central not just to issues regarding interface design, but more importantly to the design of virtual reality environments themselves. Minsky's ideas led a community of researchers to form the field of presence theory, dedicated to the study of the phenomenon. The group sought to determine the best ways of defining and measuring presence in order to inform the design of virtual reality environments and the corresponding hardware.

The history of presence research, however, is replete with definitional conflicts. Aside from the general consensus that presence concerns the sensation of being inside a virtual environment, there have been considerable disagreements over the definitions of main terms in the field. The first terminological divergence from Minsky's original term "telepresence" came in 1992 with the launch of the academic journal *Presence*. In the first issue, Thomas Sheridan (1992) reserved *telepresence* for cases of teleoperation and coined the term "virtual presence" to refer specifically to presence in virtual environments. Richard Held and Nathaniel Durlach (1992), on the other hand, used the term to refer to both cases. Later, articles in presence theory dropped "telepresence" altogether and used "presence" to refer to experiences in both virtual and actual environments.

These differences are not merely terminological, but ontological. The dominance of Held and Durlach's view in contemporary presence theory has resulted in an assumed equivalence between the experience of presence in virtual and physical environments, as in this comment: "Importantly, multisensory stimulation arises from both the physical environment as well as the mediated environment. There is no intrinsic difference in stimuli arising from the medium or from the real world—the fact that we can feel present in either one or the other depends on what becomes the dominant perception at any one time" (Ijsselsteijn and Riva, 2003, 6).

In claiming that stimuli arising in virtual and physical environments are equivalent, Wijnand Ijsselsteijn and Giuseppe Riva sideline the contrived nature of a digitally designed environment and the modes of interaction and interpretation that result from this. That is, beginning with the assumption that the two environments are intrinsically the same with regards to participants' mode of being ignores the fundamentally different forms of practice possible in each. Even when virtual environments are designed for more open-ended behavior, as is the virtual world *Second Life* (Linden Lab, 2003), we have no conclusive evidence that players are, at any point in their interaction, confused about the designed nature of the environment. This is not to say that virtual environments do not have the potential for facilitating a variety of complex forms of interactions, but that their designed nature cannot be completely ignored by players.

By not differentiating between stimuli arising from physical and virtual environments, Ijsselsteijn and Riva contribute to the conceptual confusion surrounding the terms *presence* and *immersion*. Claiming that stimuli arising from a medium are not distinguishable from those derived from the physical world makes it possible to further argue that a technological medium's properties determine the experiences that users will have in interaction with that medium. This assumption creates a number of conceptual problems for understanding the phenomenon of presence, as we shall see below.

Mel Slater is a strong proponent of this view, arguing that a high-fidelity sound system makes listeners feel as though they were listening to a live orchestra, whether or not they find the music itself engaging:

Suppose you shut your eyes and try out someone's quadraphonic sound system which is playing some music. "Wow!" you say, "that's just like being in the theatre where the orchestra is playing." That statement is a sign of presence. You then go on to say, "But the music is really uninteresting and after a few moments my mind started to drift and I lost interest." That second statement is *nothing to do with presence*. You would not conclude, because the music is uninteresting, that you did not have the illusion of being in the theatre listening to the orchestra. The first statement is about form. The second statement is about content. Presence is about form, the extent to which the unification of simulated sensory data and perceptual processing produces a coherent "place" that you are "in" and in which there may be the potential for you to act. The second statement is about content. A [virtual environment] system can be highly presence inducing, and yet have a really uninteresting, uninvolving content (just like many aspects of real life!). (Slater, 2003, 607)

The assumption here is that fidelity creates an undeniable pull on the listener's consciousness, creating a sense of presence. This conception of media technologies does not give enough importance to the key role that interpretation and agency play in creating a sense of presence. Interpretation does not need to be a conscious action. Most interactions with an environment are possible because we have an internalized knowledge of how various aspects of that environment work. When we are faced with experiences we cannot readily interpret, our mode of being becomes more critically removed and we must actively think about what we are doing (Heidegger, 1993). Our prior experience, expectations, and knowledge form a crucial part of this interpretative relation. Thus, if you have never attended a classical music concert, nor even seen a video of one, you might not be able to feel "present" in the way Slater describes when listening to the sound system. In fact, the nature of the content in conjunction with the participant's interpretative apparatus and prior lived experience is crucial for a sense of presence. While high-fidelity systems are an important part of enhancing the intensity of an experience, they cannot in themselves create a sense of presence.

As mentioned earlier, progress in the field of game studies has been hampered by a lack of agreement on the use of both *presence* and *immersion*. These two terms are sometimes used interchangeably, yet at other times are given more specific and complementary meanings (Ijsselsteijn and Riva, 2003). It is also not uncommon to find conflicting or contradictory applications by different theorists. For example, Mel Slater and Sylvia Wilbur (1997) define *immersion* as "a description of a technology that describes the extent to which the computer displays are capable of delivering an inclusive, extensive, surrounding and vivid illusion of reality to the sense of a human participant" (606).

Slater and Wilbur contrast immersion with presence, which is defined as "a state of consciousness, the (psychological) sense of being in the virtual environment" (607). Slater (2003) subsequently refines the latter distinction, describing immersion as "simply what the technology delivers from an objective point of view" and presence as "a human reaction to immersion" (1). In this view, the term *immersion* is being used to describe the affective properties of the hardware, while presence is the psychological response to this technology. Bob Witmer and Michael Singer (1998), on the other hand, view immersion as "a psychological state characterized by perceiving oneself to be enveloped by, included in, and interacting with an environment that provides a continuous stream of stimuli and experiences" (227).

While Witmer and Singer use *immersion* in the same way that Slater and Wilbur use the term *presence*, they view presence as a combination of involvement and immersion. As we shall see in the following section, this is also the more common view of immersion adopted in game studies following Brenda Laurel's (1991) and Janet Murray's (1998) uses of the term. Recent studies (Van den Hoogen, Ijsselsteijn, and De Kort, 2009; Ijsselsteijn, 2004; Rettie, 2004) coming out of presence research follow Slater's conceptualization of immersion as an objective property of the technology and presence as a psychological reaction to this technological property.

Another problem that Slater's perspective poses is a lack of distinction between involvement and presence in ergodic media like games and non-ergodic media like film and literature. On the one hand, the properties of a technology are seen as determining presence, while, on the other, the specific qualities and affordances of the medium are sidelined. The idea that one can experience presence in ergodic and non-ergodic media is now common enough in presence research that it is generally taken as a given (Witmer and Singer, 1998; Schubert and Crusius, 2002; Gysbers et al., 2004; Marsh, 2003; Lee, 2004). Consequently, presence researchers have stretched the concept of presence to account for what is essentially an entirely noetic phenomenon. Even if we argue that certain qualities of the medium and text in question afford such an experience, the phenomenon remains within the domain of subjective imagination. The scientific community initially coined the term *presence* because a new technology enabled a qualitatively different form of experience than had been possible before its inception. In this case, two forms of experience: telepresence, in the case of teleoperation of machinery in remote physical environments, and presence in the sense of inhabiting a virtual environment. The essential quality of these two experiences lies within the ability of the system to recognize and react to the user's actions and spatial location. Extending the term to cover imagined presence in works of literature, film, or free-roaming imagination sidelines the core concern: the description and exploration of a phenomenon enabled by a specific technology.

Given the essential difference between ergodic and non-ergodic media, it is crucial for a precise inquiry into the phenomenon of presence to make a distinction between simply imagining one is present in a scene and the considerably different phenomenon of having one's specific location and presence within a virtual world acknowledged by the system itself. In an imagined scene, whatever happens is simply willed by the imaginer, who usually knows that she is directing the composition and events of the scene. In the case of a game, we can think of the player as being *anchored*, via her avatar, in the game world, allowing the game's environment and entities to react to her. This aspect of games fundamentally alters how the player perceives herself within the world, and is not present in literature, films, or personal imagining. When we identify with a character in a movie or book, or imagine we are in the same room as the protagonist, we have no way of altering the course of events, no way of exerting agency. Likewise, the environments and characters represented in these media have no way of reacting to our presence, no matter how strongly we identify with them.

Presence as Transparency

In an attempt to reconcile the disagreements within the field, Matthew Lombard and Theresa Ditton (1997) surveyed the various uses of *presence* in the literature. They identified six characterizations of presence: "presence as social richness," "presence as realism," "presence as transportation," "presence as immersion," "presence as social actor within a medium," and "presence as medium as social actor." From these six conceptualizations they drew a common thread, which they used to posit a definition of *presence* as "the perceptual illusion of non-mediation." They further divided the category of "presence as immersion" into perceptual and psychological immersion. Perceptual immersion is similar to Slater and Wilbur's (1997) view of immersion as the hardware's effect on the user's senses. Psychological immersion is equivalent to the metaphor of *immersion as absorption* as used by Brown and Cairns (2004), Douglas and Hargadon (2001), Ermi and Mayra (2005), Jennett et al. (2008), Salen and Zimmerman (2003), and others.

Lombard and Ditton's definition of *presence* points to an important dimension of the phenomenon that Jay Bolter and Richard Grusin (1999) have called the "logic of transparent immediacy" (23). The transparency alluded to is that of the interface. Transparency erases the interface and offers the viewer or user as direct an experience of the represented space as possible. Techniques of transparency combine content and form to deliver "the perceptual illusion of non-mediation" (Lombard and Ditton, 1997). But, as has been argued in various areas of aesthetic inquiry, transparency is not unique to virtual environments. As mentioned earlier, Grau (2003) identifies a history of immersion in painted "spaces of illusion" dating back to ancient Roman villas:

Immersion arises when the artwork and technical apparatus, the message and medium of perception, converge into an inseparable whole. At this point of calculated "totalisation," the artwork, which is perceived as an autonomous aesthetic object, can disappear as such for a limited period of time. This is the point where being conscious of the illusion turns into unconsciousness of it. As a general rule, one can say that the principle of immersion is used to withdraw the apparatus of the medium of illusion from the perception of the observers to maximize the intensity of the message being transported. The medium becomes invisible. (349)

As Bolter and Grusin (1999) state, this logic of transparency is a salient characteristic of immersion in virtual environments. Although the transparency of the medium and text discussed by Bolter and Grusin, Grau (2003), Lombard and Ditton (1997), and others is an essential quality of immersion, it is not by itself sufficient to describe the multiple dimensions of the experiential phenomenon.

When Lombard and Ditton's conception of presence is applied to practical analysis, it fails to yield rich results. Alison McMahan's (2003) "Immersion, Engagement and Presence: A Method for Analyzing 3-D Video Games," for example, applies Lombard and Ditton's concept to an analysis of the game *Myst III: Exile* (Presto Studios, 2001). Yet McMahan's essay is little more than a descriptive listing of game features classified according to Lombard and Ditton's (1997) six categories of immersion. This results in a pigeonholing of aspects of the game without bringing out the specific affordances of *Myst III* which foster a sense of presence. When describing social realism, for instance, McMahan states that:

Players cannot see their own reflection in glass or water, or even see their own feet when they look down, but they can take rides in elevators and zeppelins and other related contraptions, can turn the pages of books, peer closely at objects, and "pick up" certain items as well as manipulate mechanical contraptions. As a result of these measures, the game has an extremely high degree of social realism, as the majority of the elements in this fantastical world conform quite closely to how things would be in our world. (81)

Claims like these can be applied to most other digital games in similar genres. Mechanical contraptions can be manipulated, vehicles operated for the sake of transportation, and book pages turned. This does not get us any closer to understanding presence in the game. When she turns to immersion, McMahan follows a common trend in game studies in conceptualizing immersion as a form of involvement, diverting it from its connotations of transportation found more commonly in presence theory.

Presence and Immersion in Game Studies

Outside of presence theory, *immersion* finds its most frequent use in the context of digital games. The application of the term, however, varies considerably: It is used to refer to experiential states as diverse as general engagement, perception of realism, addiction, suspension of disbelief, identification with game characters, and more. This plethora of meanings is understandable when it comes to industrial or popular uses of the term, but it is also common within academic game studies. Given that the phenomenon that *immersion* and *presence* have been employed to refer to is increasingly important in shaping the experience of digital games, we require a more precise approach.

As the representational power of computer graphics and audio increases, game companies have adopted *immersive* as a promotional adjective to market their games. This strategy was initially employed almost exclusively to promote photorealistic graphics, but now is also used to market other features, such as the scope of the game world, the artificial intelligence, or an engaging narrative:

Conan takes graphics in MMOs to a new level! With the latest and greatest in technology and an amazing art direction the graphics in Conan immerses you into a world as never before seen in any online fantasy universe. (Funcom, 2006)

Taking place in a massive, free-roaming city featuring five distinct interconnected neighborhoods, Need for Speed Underground 2 delivers an immersive game world where the streets are your menus. (Electronic Arts, 2006)

Half-Life sends a shock through the game industry with its combination of pounding action and continuous, immersive storytelling. (Valve Software, 2004)

The underlying assertion in these and other examples is that immersion is a positive experiential quality of games that is desirable for the consumer. At times immersion seems to be seen as something of a holy grail within the game industry because of its connection with an engagement that draws players so deeply into the game world that they feel as if they are part of it. There are echoes of cyberpunk romanticism here and perhaps an unstated ideal desire to delve, Neo-like (Wachowski and Wachowski, 1999), into a virtual reality that replaces the realm of physical existence.

This idealization of total immersion has been critiqued by Katie Salen and Eric Zimmerman (2003), who argue that too many designers share this mimetic imperative in game design. They cite François Laramée's statement as a representative example: "All forms of entertainment strive to create suspension of disbelief, a state in which the player's mind forgets that it is being subjected to entertainment and instead accepts what it perceives as reality" (Laramée, quoted in Salen and Zimmerman, 2003, 450). Salen and Zimmerman strongly oppose this view, which they call the "immersive fallacy":

The immersive fallacy is the idea that the pleasure of a media experience lies in its ability to sensually transport the participant into an illusory, simulated reality. According to the immersive fallacy, this reality is so complete that ideally the frame falls away so that the player truly believes that he or she is part of an imaginary world. (451–452)

For Salen and Zimmerman, the drive toward sensory immersion is not the most important aspect of game enjoyment and engagement. They instead argue for designers to come up with more engaging gameplay mechanics. They support their argument by citing Elena Gorfinkel's views on immersion:

The confusion in this conversation has emerged because representational strategies are conflated with the effect of immersion. Immersion itself is not tied to a replication or mimesis of reality. For example, one can get immersed in Tetris. Therefore, immersion into gameplay seems at least as important as immersion in a game's representational space. (Gorfinkel, quoted in Salen and Zimmerman, 2003, 452)

For Salen and Zimmerman, *Tetris* (Pajitnov, 1985) is a telling example of a game in which "immersion is not tied to a sensory replication of reality" (Salen and Zimmerman, 2003, 170). They rightly highlight the problems associated with equating immersion with representational mimesis and the merits of avoiding design principles based solely on the pursuit of greater realism. But this point is made at the cost of the specificity of meaning that *immersion* has accrued in discussions surrounding virtual environments. When Gorfinkel states that one can become immersed in *Tetris*'s gameplay, she is referring to the more general, pre–virtual environment sense of the word as defined by the *Oxford English Dictionary* (2003): "Absorption in some condition, action, interest, etc." In this sense, one can be just as immersed in solving a crossword puzzle as in *Half-Life 2* (Valve Software, 2004). We will call this kind of immersion *immersion as absorption*.

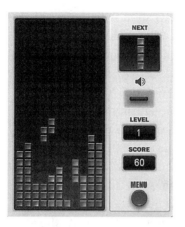

Figure 2.1
Tetris (Pajitnov, 1985) and *Half-Life 2* (Valve Software, 2004).

The problem here is that the *absorption* sense of *immersion* jettisons a history of application in the context of virtual environments within both the humanities (Murray, 1998; Ryan, 2001; Laurel, 1991) and presence theory (Steuer, 1992; Tamborini and Skalski, 2006; Ijsselsteijn, 2004; Ijsselsteijn and Riva, 2003; Waterworth and Waterworth, 2003; Slater, 2003). We will call this second use of *immersion*, which refers to the idea of being present in another place, *immersion as transportation*. Thus, a game like *Half-Life 2* presents the player not just with an engaging activity, but also with a world to be navigated. A player who assimilates this game world into their gameplay as a metaphorically habitable environment can be thought of as being *transported* to that world. This experience is made possible by the anchoring of the player to a specific location in the game world via their avatar, which the game world and its inhabitants, including other players, react to.

Games like *Tetris*, *Bejewelled* (PopCap Games, 2001), or *Eliss* (Thirion, 2009) do not afford this kind of spatial habitation. There are at least two reasons for this. First, none of these games recognizes the presence of the player within a single location in its environment. In each case the player controls objects (blocks, gems, planets) without being embodied in any single in-game entity. Second, the game environment is represented in its totality on one screen; there is no element of continuous spatial navigation. This is also true of strategy games in which the player controls multiple miniatures or a collective unit such as a nation without being digitally embodied in the world through an avatar. By engaging with the game world as a *map* rather than as a spatial environment, the player remains in conceptual rather than inhabited space.

In emphasizing immersion as absorption in a game like *Tetris*, Gorfinkel, Salen, and Zimmerman sideline the importance of spatiality as a defining feature of the phenomenon to which *immersion* has been used to refer in the context of virtual environments. Marie-Laure Ryan (2001) has argued that the representation of space, whether it is internally generated or graphically displayed, is one of the key features of the subjective experience of immersion: "For a text to be immersive then it must create a space to which the reader, spectator, or player can relate and it must populate this space with individuated objects. ... For immersion to take place, the text must offer an expanse to be immersed within, and this expanse, in a blatantly mixed metaphor, is not an ocean but a textual world" (90).

It is counterproductive for a field to use a term like *immersion* in its earlier, more general sense when it has since accrued specific meaning within its discursive domain. This is particularly true when theorists do not clarify which of the two meanings they using in a specific study. The blurring of difference between the two applications of *immersion* as absorption and as transportation also obscures the fact that game environments enable qualitatively different forms of engagement from other media. This reduces, rather than develops, the critical vocabulary available to game studies.

Unfortunately, the confusion surrounding *immersion* is pervasive in the literature and discussions accumulating within the field. Jon Dovey and Helen Kennedy (2006), for example, use the terms *immersion* and *engagement* interchangeably to refer to the absorbing qualities of digital games:

This quality of immersion or engagement within the game world may account for the ways in which a sense of time or physical discomfort may recede as the player's skill develops. This is a critical aspect of the unique time economy that characterizes computer gameplay. It is entirely commonplace that gameplay experience seems to lie outside of day to day clock time—we sit down to play and discover that hours have passed in what seemed like minutes. (8)

The engaging qualities mentioned by Dovey and Kennedy are neither specific to nor guaranteed by game environments. This sense of immersion as absorption makes the term as readily applicable to gardening or cooking as it does to game environments. While there is nothing wrong with this use of the term in itself, it undermines the more specific sense of transportation that is crucial when discussing game environments.

Building on Victor Nell's (1988) work on the psychology of reading, Jane Douglas and Andrew Hargadon (2001) make a distinction between immersion and engagement based on the critical stance of the reader or player (they do not make a distinction between the two). In cases of immersion, "perceptions, reactions and interactions all take place within the text's frame, which itself usually suggests a single schema and a few definite scripts for highly directed interaction" (Douglas and Hargadon, 2001, 152). They contrast this with a state of engagement, which is a more distanced and critical mode of experience required by more complex texts. Engagement calls upon a more conscious interpretative effort that results in a discontinuous interaction with the text: "Conversely, in what we might term the 'engaged affective experience,' contradictory schemas or elements that defy conventional schemas tend to disrupt readers' immersion in a text, obliging them to assume an extra-textual perspective on the text itself, as well as on the schemas that have shaped it and the scripts operating within it" (152).

This conceptualization acknowledges a continuum of experience moving from conscious attention to unconscious involvement that is important in understanding the process of engagement with games. But the argument becomes challenging to sustain when the authors sideline the considerably different qualities of literary works and games, and consequently the forms of engagement each affords: "Ironically the reader paging through Balzac, Dickens or for that matter, Judith Krantz, has entered into the same immersive state, enjoying the same high continuous cognitive load as the runty kid firing fixedly away at *Space Invaders*" (Douglas and Hargadon, 2001, 157). Book readers might imagine themselves within the space world described by a literary work, but that world does not recognize them. On the other hand, game environments afford extranoetic habitation by recognizing and reacting to the presence of the player. Books also do not provide readers with the possibility of actually (not imaginatively) acting within the worlds they describe. Thus, Douglas and Hargadon's exposition of immersion remains limited to different forms of involvement without capturing the distinctive qualities of the phenomenon that immersion as transportation points to.

Emily Brown and Paul Cairns (2004) similarly use the term *immersion* synonymously with *absorption*, describing three degrees of immersion: engagement, engrossment, and total immersion. *Engagement* describes the player's interest in engaging with the game. This is the most basic form of involvement, which Brown and Cairns relate to the desire and ability to interact with a game. *Engrossment* describes a deeper level of involvement characterized by emotional attachment to a game. They also equate this degree of emotional affect with a loss of awareness of self and surroundings. Although affect and loss of awareness of surroundings can and do mingle in game-related and everyday experiences, they do not seem to be intrinsically tied together, and Brown and Cairns do not provide us with an explanation for why they might be. Finally, the authors identify *total immersion* as the most intense form of involvement, and argue that this state is synonymous with presence.

Brown and Cairns's emphasis on increasing degrees of involvement is certainly a useful acknowledgment of the varying intensities of engagement that games accommodate. The problem with this conceptualization of immersion is that it mixes the two metaphors we are exploring. At the first two levels, immersion is seen as the same as absorption, but at the third and most intense stage of total immersion, the metaphor switches to immersion as transportation. On the one hand, this work is productive in identifying increasing levels of involvement as prerequisites for a sense of immersion as transportation, but on the other, Brown and Cairns's formulation is problematic because it exacerbates the definitional confusion surrounding *immersion* that the authors had set out to resolve.

A later paper by Cairns et al. (2006) applies this conceptualization of immersion, as "the degree of involvement with a computer game" (para 8), to an experiment designed to quantify degrees of immersion. The experiment proposed two hypotheses: first, that the level of immersion during gameplay would be higher than in a control task of repeatedly clicking a button with a mouse; second, that participants would perform worse in a tangram task following gameplay than following the control task. The latter hypothesis was based on the premise that "if a person becomes present in some alternative game world, then there may be some measurable effect on their 'return' to the real world" (para 8). This is another example of a mixed application of the terms. Although the researchers set out to explore immersion as absorption, the assumptions underlying their hypotheses are tied to a sense of presence in a virtual world inherent in the immersion-as-transportation metaphor. This is problematic because it is not clear whether the conclusions derived from the experiment relate to the absorption or transportation metaphors of immersion. A second problem with the study is that, although it claimed to explore a phenomenon specific to games, the impairment of subsequent task completion would be equally relevant to that caused by any deeply involving cognitive task. Such impairment has little bearing on the specific experience of game playing, or of shifting experiential worlds, but is a function of cognitive readjustment and fatigue. We have all had the experience of being distracted for a brief period after any cognitively intense activity, from reading a book to watching a television show. This overgeneralized view of immersion makes quantifiable measurement difficult, if not impossible, until the various forms of involvement are explored, conceptualized, and tested.

Laura Ermi and Frans Mayra (2005) address what they feel is a lack in Brown and Cairns's (2004) study by providing a multidimensional model for immersion based on three modes of immersion: sensory, challenge-based, and imaginative. Sensory immersion relates to engagement with the representational, audiovisual layer of games. Challenge-based immersion addresses the employment of both mental and motor skills in overcoming challenges presented by the game. Imaginative immersion seems to be a catchall category that encompasses everything from identification with a character to engagement in the narrative and game world.

Ermi and Mayra's "gameplay experience model" makes an important contribution to the field by acknowledging that involvement is a multidimensional phenomenon. It also, importantly, emphasizes the difference between involvement in games and involvement in other media. Ermi and Mayra are also consistent in their application of the notion of immersion as absorption, which they derived from a series of interviews with children. There is, of course, no reason why a term cannot be used in a particular sense if it is adequately defined, but there is a difficulty in attempting to empirically establish the meaning of a term like *immersion* by asking the general populace what it means. The general populace will usually not be aware of the theoretical connotations of the term and will thus give a more generic perspective on it. This is a problem when the term is then used in the field in a more colloquial sense that clashes with the more academically technical one.

The Four Challenges

Theoretical concepts and metaphors accumulate a specific meaning that does not necessarily translate outside of the academic context in which they are used. If we want to describe the more general forms of involvement in games, we can use terms that do not have the theoretical associations of *immersion*, such as *engagement, absorption,* or *involvement.* Ermi and Mayra's (2005) study can be more usefully couched in the context of such terms without invoking the confusion over how to define *immersion.* What is problematic, from a theoretical perspective, is when investigations of involvement, like those cited above (Jennett et al., 2008; Ermi and Mayra, 2005; Brown and Cairns, 2004; Cairns et al., 2006), refer to the discussion of presence and immersion (in the sense of virtual environment habitation) with an aim of contributing to a clearer understanding of these concepts, and then offer a view of immersion in its more general sense of involvement. The terms might be the same, but the experiential phenomenon they are investigating is not.

Conducting studies without acknowledging that the central term we are investigating has a dual meaning in the colloquial and academic domains will inevitably lead to difficulties in making progress. This is not to say that we do not need to study immersion as absorption or involvement. Quite the opposite: It is precisely with involvement that we must begin our investigations, but we would be wise to avoid confusion and call it what it is—involvement.

In this chapter we have thus identified four key challenges to gaining a clear understanding of immersion:

1. *Immersion as absorption versus immersion as transportation*   There is a lack of consensus on the use of *immersion* to refer to either general involvement in a medium (Salen and Zimmerman, 2003; Jennett et al., 2008; Ermi and Mayra, 2005) or to the sense of being transported to another reality (Murray, 1998; Laurel, 1991; Carr, 2006). This is particularly problematic when researchers do not clarify which one of these terms they are using

or when they oscillate between the two within the same study (Brown and Cairns, 2004; Cairns et al., 2006).

2. *Immersion in non-ergodic media*   For a precise formulation of both immersion as absorption and immersion as transportation, we need to acknowledge the specificities of the medium in question. In this case, immersion in ergodic and immersion in non-ergodic media are simply not the same thing. The challenge of addressing a complex and preconscious phenomenon such as immersion as transportation is increased considerably if we try to extend the concept to multiple media with considerably varied qualities and affordances for engagement.

3. *Technological determinism*   Although the specifics of the medium are crucial for our understanding of the experiences they afford, we should avoid seeing such experiences as being determined by the qualities of the technology. A bigger screen and a higher fidelity of representation, for example, might make it easier for users to focus and to keep their attention on the representation, but this does not necessarily mean that users will feel more present in the environment portrayed.

4. *Monolithic perspectives on immersion*   The principal reason for (3) is that whether *immersion* is defined as absorption or as transportation, both are made up of a number of experiential phenomena rather than being a single experience we can discover and measure. The various forms of experience that make up involvement need to be considered on a continuum of attentional intensity rather than as a binary, on/off switch.

Whichever term we use, whether we try to clarify the use of *presence* or *immersion*, or propose a new metaphor, as will be done here, these challenges need to be addressed before progress can be made in further exploring the phenomenon.

Toward a Solution

Virtual environments offer a particular form of mediated experience that was not previously possible. Two terms have been formulated in different disciplines to articulate this experiential phenomenon. Technologists, media psychologists, and human-computer interaction researchers, among others, refer to this experience as *presence*, while humanists and, later, social scientists adopted the metaphor of *immersion*. The terms have both suffered from varied application and their use has generated a number of debates, primarily relating to the breadth of experiential domains they cover.

This is not to say that we cannot make advances in exploring this phenomenon, but we need to be realistic about the limitations underlying such an endeavor. The first step to understanding our sense of inhabiting virtual environments is therefore the establishment of clear and precise terms that take into consideration the specifics of the medium without making normative statements about how such specific characteristics determine experience. To do this, we need to be critical of the implications and assumptions written into the metaphors we use.

Any treatment of the experientially complex phenomenon of presence in virtual environments must first consider the structure of its key prerequisite: involvement. We cannot feel present anywhere without first directing our attention toward and becoming involved with the environment. We thus need a model for understanding player involvement in virtual game environments. Once the various forms of involvement are outlined, we can proceed to examine how these combine and interact in consciousness to create a sense of presence. The basic assumption made here is that whatever we decide to call this phenomenon, it will not be a single, monolithic form of experience, but will emerge from the combination of these forms of involvement. In order to limit the considerable list of unknowns related to examining such experiential phenomena, we must focus on limiting the scope of our inquiry to the specific affordances of virtual game environments. The rest of this book will thus present a model for understanding player involvement in virtual game environments.

3 The Player Involvement Model

Overview of the Model

The previous chapter identified the key challenges to a better understanding of immersion and presence. To address these challenges, I will first establish a better understanding of player involvement. Involvement is a prerequisite to the experience of higher-order cognitive processes such as presence or immersion in much the same way that attention is a prerequisite of involvement. It therefore makes sense to establish a thorough model of involvement before going on to attempt a formulation of what is essentially a preconscious experiential phenomenon that combines multiple dimensions of involvement.

Chapter 2 described the scarcity of comprehensive conceptual frameworks that can be employed to understand the multiple facets of player involvement and, consequently, immersion in digital games and virtual worlds. Having reached this conclusion in my initial research, I felt it was important not only to relate the theoretical works on the subject to observations drawn from my own experience in games, but to observe the views of other players through personal participation in the same game world as the participants and, more specifically, by way of a series of focus groups and in-depth interviews with them. This chapter gives a brief outline of the research process that contributed to the creation of the player involvement model along with the theoretical considerations that factored into its structure and an overview of the model's constituent parts.

The Research Process

The player involvement model presented in this chapter is the product of three years of qualitative research, focusing primarily on two popular and contrasting massively multiplayer on-line games (MMOGs): *World of Warcraft* (Blizzard Entertainment, 2004), the prototypical massively multiplayer on-line role-playing game (MMORPG); and *Planetside* (Sony Online Entertainment, 2003), a massively multiplayer on-line first-person shooter (MMOFPS). A qualitative perspective was adopted due to the intensely experiential and richly varied nature of the object of inquiry, following other researchers of virtual worlds such as T. L. Taylor (2006), Constance Steinkuehler (2005), and Lisbeth Klastrup (2004).

The primary research methods were extended personal participation in the two selected game environments, numerous focus groups, and in-depth interviews with twenty-five experienced players recruited in-game. These techniques were supported by discussions with academics studying virtual worlds, participation in on-line forums, monitoring of player blogs, and viewing player-made machinima.

The interviews, which provided the core data for analysis, were semistructured and extended over two sessions, each up to two hours long. Each interview took place within the virtual environment of the game about which the participant was being interviewed, in order to further stimulate their memory and to bring their relationship to the game world to the fore. The interviews followed a loosely structured schedule with a focus always placed on experiences related to immersion.

The data was coded and analyzed between interviews, with the interview schedule updated in the intervening period to reflect input from the participants. Over time, a number of strong themes emerged from the interviews, and the player involvement model is based on these salient features of the data.

Structure of the Model

When I analyzed the qualitative research data gathered for this project, it became clear that it was important to make a distinction between aspects of a game which engaged players in the moment of playing from aspects that attracted players to the game initially and kept them returning to the game over time. I refer to these aspects as *micro-involvement* and *macro-involvement* respectively. Let us look at a brief example of what each of these covers.

In my own experience, I find that what involves me deeply in *Empire: Total War* (Creative Assembly, 2009) are its on-line multiplayer battles. Here you select and control an eighteenth-century army in a real-time battle against a human opponent. A crucial part of the game is building one's army with the limited funds set by the host of the battle. Next, all players deploy their armies in a stipulated zone and the game starts in earnest. The plans and choices I make during the game, my control of the individual units on the battlefield, and my attempts at outmaneuvering my opponent are all examples of aspects of the game which involve me deeply during gameplay. These aspects of involvement in the moment of play are all part of the micro-involvement phase.

When I turn off the PC and go to bed, I inevitably start thinking of different army configurations and other ways to use certain units. Like many gamers, I run through what happened in my recent battles and why certain tactics failed and others succeeded. In *Empire: Total War*, the campaign mode starts in 1700 and allows players to manage one of the many factions in the game in a bid to become the strongest empire in the world at the time. Plans for furthering one's conquest and dealing with deteriorating relations with neighboring powers are formulated not just during gameplay but also during off-line thinking about the game, such as when one is riding the train or in other situations which do not require one's full attention. I refer to this form of ongoing motivation to interact with the game and the off-line thinking that fuels it as *macro-involvement*.

The player involvement model identifies six dimensions of involvement, each considered relative to two temporal *phases*: the macro and the micro. The six *dimensions* correspond to the clusters of emphasis derived from analysis of the research data. The dimensions are experienced not in isolation but always in relation to each other, the separation being made here for the sake of analysis. Dimensions are experienced unconsciously during the interpretative and communicative process and therefore play an important role in noticing and directing attention toward aspects of a given reality. In the case of the player involvement model, this reality is made up primarily of stimuli originating from the game

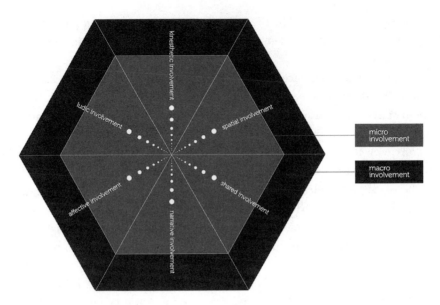

Figure 3.1
The player involvement model.

environment, but at times the stimuli originate from the surrounding physical environment.

The dimensions should be seen as layered and transparent in nature. This means that one dimension influences how another is perceived and interacted with. The dimensions are transparent in that their layering does not occlude what lies beneath, but changes the perception of both. The dimensions of the player involvement model similarly combine in the game-playing experience, with the inclusion or exclusion of a dimension affecting how others operate.

The six dimensions of the player involvement model are *kinesthetic involvement*, *spatial involvement*, *shared involvement*, *narrative involvement*, *affective involvement*, and *ludic involvement*. The following chapters will elaborate at length on the macro and micro phases of each of these dimensions, but this chapter will give a brief outline of each.

Macro-involvement

As Karolien Poels (2009) notes, research on game experiences has focused primarily on the moment of gameplay. There are, however, a number of factors that shape the player's opinion and disposition toward the game that derive from thoughts, plans, feelings, and expectations both prior to and following the game experience. Examples of these out-of-game forms of involvement were common in the accounts given by participants in my research, and included the feeling of belonging to a close-knit community, the formation of strategies and plans that could be tried in upcoming sessions, interest in continuing to unravel a game narrative or exploring a newly discovered part of the game world, the ambition to develop one's abilities and outperform others, or simply the desire to feel surrounded by a beautifully rendered world of a desired setting. This off-line involvement is described by the macro phase of the model. It concerns issues of motivations and sustained engagement with digital games through the long-term (as opposed to immediate) aspects of the six dimensions of involvement that make up the model.

Poels et al. (2010) conducted two focus group studies exclusively concerning what they call "postgame experiences." Their findings were very much in line with my own. Their participants reported a variety of forms of postgame involvement that could be mapped onto the dimensions of the player involvement model. Although Poels et al. were researching a broader range of experiences than those directly related to involvement, their findings corroborate the importance of not limiting the analysis of involvement to the game-playing session itself.

Postgame experiences can range, for example, from the sense of accomplishment derived from completing an elusive game goal (ludic involvement) to the satisfying feeling of recalling impressive feats of avatar control (kinesthetic involvement) or a sense of inner peace following travels in aesthetically moving surroundings (spatial and affective involvement).

Pregame experiences are also important to consider because they give a context for the individual game-playing session. It is harder to model this form of involvement since it is guided less by the formal aspects of the game than is the case in the micro phase, but this is not to say that the formal characteristics of games are irrelevant for macro-involvement. As we will see in the exposition of the dimensions in upcoming chapters, off-line involvement can also be about coming up with winning strategies or planning how to develop one's character as she levels up (ludic involvement), wondering how the designed narrative surrounding a particular character will develop (narrative involvement), or working on a community Web site for one's clan or guild (shared involvement).

Both postgame and pregame experiences are covered by the macro phase of the player involvement model. Since this phase exists largely as a contextualization of the actual moment of gameplay, in the following chapters I will address more general issues surrounding involvement in the particular dimension in question, as a way to frame the discussion of actual play and the corresponding micro-involvement in the rest of each chapter.

The macro phase of the player involvement model addresses longer-term motivations as well as off-line thinking and activities that keep players returning to a game. Examples of this were common in the responses of participants in my research. They often commented on the plans and strategies they formed during work hours, in conversation with others, or through on-line resources and community sites. Other participants described the construction of stories featuring their in-game characters, and the majority expressed the importance of persisting social bonds to their prolonged engagement with a game.

Micro-involvement: In Ludo Res

We now shift our attention from the broader motivations that attract players to games to the moment-by-moment engagement of gameplay described by the micro-involvement phase of the player involvement model. A crucial first step for forming a conceptual toolkit that will help in analyzing and discussing game experience is to make a distinction between the general direction of attention toward a medium and the form of active involvement prevalent during gameplay.

All forms of representational media require the investment of attention in order to interpret them. Without attention there can be no involvement. The term *attention* generally refers to the concentration of mental resources toward some particular stimulus or stimuli. This involves an assortment of skills, processes, and cognitive states that interact with each other and with other brain processes (Fan et al., 2002). Attention underlies everything that we do, and plays a crucial role in perception, thinking, learning, and performance. When the brain carries out activities simultaneously, the coordination of their execution also involves attention.

Most of the time, we are not aware of the way in which attention affects our performance or behavior. It becomes more apparent when, for example, we are trying to comprehend complex information, learn a new task, or engage in activities that are unfamiliar to us. In such situations, the information required to solve the task or manage the situation can be greater than what our attentional capacity system can handle (Baddeley and Hitch, 1994; Sturm and Willmes, 2001). As a result of this innate capacity limit of the human information processing system, our attentional resources are allocated to those aspects of a task most relevant to us at that particular time. Learning a particular task requires the ability to attend to the relevant stimuli, and as the information is transferred to long-term memory storage and is learned, the behavior required by the task becomes internalized and automatic. The learned task will thus require less attentional resources, freeing resources that now can be allocated to new tasks.

Although attention is a key prerequisite for involvement, it does not adequately describe the experience. It is helpful, therefore, to differentiate between general attention directed toward a medium and the active input from the player. Watching a movie does not require the same kind of involvement as game playing or navigation of a virtual environment, and treating them as experiential equivalents ignores the specific qualities of each experience. This situation has, at times, caused difficulties within game scholarship when analytical frameworks from other disciplines have been imported without modification to the study of games. Espen Aarseth's (1997) *Cybertext* focuses on the different relationships between the readers of a literary work or film viewers and what he calls the "operator" of a cybertext. The cybertext is characterized by *ergodicity*: the operator, or in our case the player, must provide active input in order for the text to come into being. The player reconfigures the constitution of the text through her input. Ergodicity thus expresses the active participation of the player within the cybernetic circuit (Dovey and Kennedy, 2006) that is formed by the game's hardware, the representational layer, and the underlying rules and environmental properties.

Ergodic forms of engagement, such as gameplay, are not limited to direct input. The effort implicit in the ergodic is first and foremost a *disposition* and readiness to act, not merely the action of pressing a button or pulling a joystick. For example, one of the pleasures of strategy games is the mental effort required to come up with a particularly brilliant plan or move. During very involving games, these periods of seeming inactivity can be long, but it would make little sense to label such periods as not

Figure 3.2
Covering a street in *Red Orchestra* (Tripwire Interactive, 2006).

being part of the game or as non-ergodic, since for these members of the game family, this is exactly what the game is all about.

This consideration is also applicable to action games. In *Red Orchestra* (Tripwire Interactive, 2006), a first-person shooter game, for example, a player is lying on the floor of a three-story ruined building, covering a street with a sniper rifle. There are no enemies in sight, but the sniper expects them to emerge in the near future as the street leads to one of the main game objectives on the map. Long minutes of inactivity result from such a wait, yet the sniper's job is often defined by this sort of patient waiting. Although there is no direct input on the part of the player, the readiness to act requires her to stay alert. At any second someone might emerge around that street corner, and the sniper must be ready to deal with him, or the fruits of her labor will go to waste.

Consulting an in-game map and planning a perilous journey from Ratchet to Feathermoon in *World of Warcraft's* (Blizzard Entertainment, 2004) Azeroth is also a significant form of ergodicity. Even though no action is apparent on the screen, game-related effort is still being invested through the player's thoughts. Game involvement is indicated not simply by the direct input of the player or the display of such an action on the screen, but by the player's *cognitive* effort, which is not necessarily registered as a form of input. Planning a move in a strategy game clearly requires effort and is thus an important aspect of ergodicity, as is the readiness to act discussed in the *Red Orchestra* example above.

The Dimensions of the Player Involvement Model

Kinesthetic Involvement

Kinesthetic involvement relates to all modes of avatar or game piece control in virtual environments, ranging from learning controls to the fluency of internalized movement. This dimension of involvement requires more conscious attention when the controls make themselves present, either because the player hasn't fully mastered them or because a situation demands a complex sequence of actions that are challenging to the player. The freedom of action allowed and the difficulty of the learning curve of the controls involved have a major influence on the player's involvement in the game environment. Kinesthetic involvement is discussed in chapter 4.

Spatial Involvement

The spatial involvement dimension, which is the focus of chapter 5, concerns players' engagement with the spatial qualities of a virtual environment in terms of spatial control, navigation, and exploration. It accounts for the process of internalizing game spaces that is a powerful factor in engaging players and giving them the sense that they are inhabiting a place, rather than merely perceiving a representation of space.

Shared Involvement

Shared involvement, the subject of chapter 6, deals with the engagement derived from players' awareness of and interaction with other agents in a game environment. These agents can be human- or computer-controlled, and the interactions can be thought of in terms of cohabitation, cooperation, and competition. Shared involvement thus encompasses all aspects relating to being with other entities in a common environment, ranging from making collaborative battle strategies to discussing guild politics or simply being aware of the fact that actions are occurring in a social context.

Narrative Involvement

Narrative involvement refers to engagement with story elements that have been written into a game as well as those that emerge from players' interaction with the game. It addresses two interrelated dimensions of narrative in games: the narrative that is scripted into the game and the narrative that is generated from the ongoing interaction with the game world, its embedded objects and inhabitants, and the events that occur there. Narrative involvement is discussed in chapter 7.

Affective Involvement

The affective involvement dimension encompasses various forms of emotional engagement. Emotional engagement can range from the calming sensation of coming across an aesthetically pleasing scene to the adrenaline rush of an on-line competitive first-person-shooter round to the uncanny effect of an eerie episode in an action-horror game. This dimension, which is the focus of chapter 8, accounts for the rhetorical strategies of affect that are either purposefully designed into the game or precipitated by an individual player's interpretation of in-game events and interactions with other players.

Ludic Involvement

Chapter 9 discusses the ludic involvement dimension, which expresses players' engagement with the choices made in the game and the repercussions of those choices. These choices can be directed toward a goal stipulated by the game, established by a player, or decided by a community of players. Decisions can also be made on the spur of the moment without relation to any overarching goal. Seasoned game players understand that well-balanced game systems emphasize the opportunity cost of any particular action taken. Without repercussions, actions lose their meaning.

Applying the Model

Applying the player involvement model to practical analysis does not require all the dimensions to be equally relevant to a specific game; for example, the intensity and complexity of spatial involvement in *The Elder Scrolls IV: Oblivion* (Bethesda Softworks, 2006) or *Half-Life 2* (Valve Software, 2004) far surpass that found in *Pong* (Atari, 1972). This does not mean that space does not play a role in a game like *Pong*, but that the potential breadth of involvement with this dimension of the game is severely limited when compared to game environments that at times represent whole regions in minute detail.

The outer edge of each triangular segment in the model represents full attentional resources directed toward that dimension of involvement. In this state, players will be attending primarily to that one dimension. The narrowing triangles of each dimension in the micro phase represent a process of internalization, where a move toward the center requires incrementally less attentional resources directed toward that dimension of involvement. With more attentional resources freed, players will attend to multiple dimensions simultaneously. The further toward the center players move, the more dimensions may be simultaneously attended to. The direction of attention tends to change frequently and fluidly, with multiple dimensions being combined and recombined as involvement proceeds.

In the kinesthetic involvement dimension, conscious attention is generally dedicated to learning the controls of the game during a player's early sessions of playing. This includes following on-screen instructions relating to avatar control or looking up and reassigning keys and buttons to the desired controls and then testing how these feel in the game. As players find the control setup that feels most intuitive to them or get used to the one supplied by the game, their conscious attention moves away from the basic controls to other aspects of the game. When a particularly demanding maneuver is required, conscious attention toward controls might resurface until the maneuver is learned, and so on. Since, as discussed above, humans have a limited attentional capacity, devoting more conscious attention to one of the dimensions leaves less that can be invested in others. As players internalize the controls, they can devote more attentional resources to understanding the layout of the city they are in, or to making plans for developing their character's attributes, and so on. In such cases, we can say that the player has "internalized" the controls: she has reached a level of kinesthetic involvement that requires little or no conscious effort.

As the player involvement model deals with an intensely subjective experiential phenomenon, there is a constant blending of dimensions and a shift from conscious to internalized attention directed to each dimension and cluster of dimensions. The challenge in building a foundation and a register to convey such a dynamic phenomenon in a diagrammatic and textually descriptive form is that we have to keep in mind the constantly fluctuating nature of experience. The model proposed here thus has a modular structure, with dimensions combining across the diagram in a fluid manner during gameplay. It is also scalable, in the sense that each of the dimensions represents a broad category of experience that can be further analyzed and fleshed out with constituent components. Each dimension will be described in more detail in the coming chapters.

General Considerations: The "Magic Circle"

The popular game studies concept of the magic circle (Huizinga, 1955) has not been included in the model. Because this concept is used widely within game studies and has an important bearing on the nature of the game experience, it is worth a brief digression to address the concept and to explain why I have excluded it from the model.

Initially coined by Huizinga (1955) in *Homo Ludens*, the concept of the magic circle has been widely adopted by game studies theorists (for example, Salen and Zimmerman, 2003; Juul, 2005) to articulate the spatial, temporal, and psychological boundary between games and the real world. As Huizinga (1955) describes the magic circle:

All play moves and has its being within a play-ground marked off beforehand either materially or ideally, deliberately or as a matter of course. ... The arena, the card-table, the magic circle, the temple, the stage, the screen, the tennis court, the court of justice, etc., are all in form and function play-grounds, i.e., forbidden spots, isolated, hedged round, hallowed, within which special rules obtain. All are temporary worlds within the ordinary world, dedicated to the performance of an act apart. (10)

Huizinga emphasizes this notion of play as "an act apart" to the extent that he describes it as a necessary condition for play to occur. He conceives of play as a "stepping out of real life into a temporary sphere of activity with a disposition all of its own" (9). According to Huizinga, all types of play, whether engaged in by humans or animals, have some form of rules, and it is the adherence to and upholding of these rules that structures and sustains the magic circle (12).

Although Huizinga sees play as separate from the real, his principal argument rests on proving that the element of play pervades (and even precedes) all aspects of human culture. The apartness of play is the apart-ness of ritual, which, Huizinga points out, shares all of the characteristics of play: "Formally speaking, there is no distinction whatever between marking out a space for a sacred purpose and marking it out for purposes of sheer play. The turf, the tennis court, the chess board and pavement-hopscotch cannot be distinguished from the temple or the magic circle" (20).

In *Rules of Play*, Salen and Zimmerman (2003) have adopted the concept of the magic circle to discuss the relationship between games and the "real world":

> The fact that the magic circle is just that—a circle—is an important feature of this concept. As a closed circle, the space it circumscribes is enclosed and separate from the real world. ... Within the magic circle, special meanings accrue and cluster around objects and behaviors. In effect, a new reality is created, defined by the rules of the game and inhabited by its players. (95–96)

Salen and Zimmerman emphasize the importance of the bounded nature of games by comparing idle toying with an object, what Caillois (1962) has referred to as *paidia*, with the formal rule-based activity, called *ludus*, of a game such as tic-tac-toe. Free play thus becomes game when the structured frame of the magic circle is imposed upon it. Salen and Zimmerman go on to argue that the magic circle surrounding games can be either open or closed, depending on the perspective, or "schema," one views them from. Games can be viewed as a system made up of rules, as a form of play activity, and as a form of culture. In the first case, games are considered as closed systems completely separate from the external world. In the second case, they can be both open and closed, since this depends upon whether or not we bracket the gameplay experience from the rest of the player's lived history. Finally, games as culture are open systems with a permeable boundary.

Salen and Zimmerman sideline the central point of Huizinga's work when they argue for an unbounded perspective on the cultural schema of games. Proving that cultural constructions are playlike and thus set aside from ordinary life is exactly Huizinga's central argument. Since the concept of the magic circle is at the heart of Huizinga's perspective, one cannot adopt it without also taking on its user's principal argument. The confusion is compounded by the fact that Salen and Zimmerman seem to be using Huizinga in a positive manner, while at the same time going against the main thrust of his argument without forwarding a coherent critique of it. Once we adopt the term, we also take on the ontology that places a distinct division between reality/seriousness/utility and play/nonseriousness/gratuitousness (Ehrmann, 1968).

Ehrmann (1968) criticizes Huizinga for conceiving of "ordinary life" or "reality" as a stable entity that can be compared, contrasted, and measured against play. Huizinga takes for granted the existence of a "reality," perpetually escorted by the hesitant presence of quotation marks, that can, in some unspecified manner, be divorced from culture and/or play. But, as Ehrmann argues, there is no reality outside of the culture that constructs it:

The problem of play is therefore not *linked* to the problem of "reality," itself linked to the problem of culture. It is one and the same problem. In seeking a solution it would be methodologically unsound to proceed as if play were a variation, a commentary *on*, an interpretation, or a reproduction *of* reality. To pretend that play is mimesis would suppose the problem solved before it had even been formulated. (33–34)

Reality cannot be bracketed by closed or open circles, even if we could argue that such bracketing is logically possible. Reality does not *contain* play; like any other sociocultural construction, play is an intractable manifestation of reality. A consideration of games—whether from the perspective of the game as an object or as an activity, or the game's role in the wider community—*is* a consideration of reality. As Taylor (2006) has argued, such a perspective ignores the grounded analysis of these objects and activities while sidelining the fact that they are very much part and parcel of everyday reality.

Separation in Space
Jesper Juul (2005) also adopts the concept of the magic circle, but he differentiates between its status in the context of what he calls physical games, like football or tennis, and in digital games. He applies the magic circle in a more specific formal capacity in terms of game space. According to Juul, physical games and board games take place in a space which "is a subset of the larger world, and a magic circle delineates the bounds of the game" (164). The boundary can be made up of spatial perimeters and is often also defined temporally. The game can be limited to a specific area such as a tennis court or a fencing piste, or woven into the everyday world, as in live-action role-playing games (LARPs), treasure hunts, and other forms of pervasive gaming. Here the spatial perimeter is less defined than the temporal one. The spatial and temporal boundaries of the magic circle in physical games are upheld by a social agreement clarifying the

interpretation and validation of actions, utterances, and outcomes; in other words, the rules.

But in the case of digital games, where is the fabled circle? Juul (2005) traces the magic circle of digital games through the hardware devices that enable their representation: "the magic circle is quite well defined since a video game only takes place on the screen and using the input devices (mouse, keyboard, controllers) rather than in the rest of the world; hence there is no "ball" that can be out of bounds" (164–165).

He goes on to compare the magic circle in physical games with that in digital games, based on the spatial qualities of each. In physical games, the magic circle separates real-world space from game space, while in the case of digital games the magic circle separates the fictional world of the game from what Juul calls "the space of a game." Juul bases this claim on an assumption that "the space of a game is *part of* the world in which it is played, but the space of a fiction is *outside* the world from which it is created" (164, his italics). In the case of digital games, the magic circle's function as a marker where rules apply loses its analytical relevance. In physical games, the distinction is needed because the game rules are upheld socially, and actions that take place within the marked area of the game are interpreted differently from actions outside that area. On the other hand, in most digital games the distinction is void because the only on-screen space that one can act in is the navigable space of the virtual environment. The stadium stands in *FIFA 2009* (EA Sports, 2008) or the space outside the combat area in *Battlefield 1942* (Digital Illusions, 2002) cannot be traversed; they are merely a representational backdrop. The role of the magic circle as spatial marker is thus redundant when applied to digital games.

Psychological Separation

More problematically, the concept of the magic circle has also been applied to the experiential dimension of gameplay. Within game studies it is often taken as a given that gameplay involves entering a particular experiential mode that was described by Bernard Suits (1978) as the "lusory attitude." The lusory attitude is closely tied to the notion of the magic circle because it is similarly built on the assumption that players voluntarily step into an attitude which is apart from ordinary life. As Suits describes this experiential mode, which occurs only during game playing: "The attitude of the game player must be an element in game playing because there has to be an explanation of that curious state of affairs wherein one adopts rules which require one to employ worse rather than better means to reach an end" (52).

According to Suits, then, we know that players are engaged in a game when they purposefully choose to engage with artificial constraints defined by the rules in order to attain a specified goal. In golf, for example, the most efficient means of sinking the ball would be to pick it up, walk over to the hole in question, and simply place the ball in it. Using a metal club to try to get the ball in the hole is an inefficient means of achieving the same goal. But this cornerstone of Suits's definition does not hold for digital games. A number of members of the game family that are simulated on a machine—that is, digital games—do not allow players to take the types of shortcuts that Suits describes. In fact, in a game of digital golf, players have no choice but to follow the rules encoded into the game and thus follow the inefficient, in the sense of rule-restricted, course of action available to them, which Suits ascribes to work. There is no need for players to make an effort to follow the rules, since the rules (or at least some of them) are coded into the game and thus are upheld by the machine. Suits's conceptualization of games thus only captures some members of the game family and does not, for example, account adequately for digital games.

Suits's notion of the lusory attitude as a defining element of games creates a problematically circular argument, essentially claiming that games are activities which require a lusory attitude and that the lusory attitude is an experience which occurs when playing a game. If we had to follow Suits's logic, the player's inability in a number of digital games, particularly single-player ones, to voluntarily adopt inefficient means in playing them means that we cannot enter into a lusory attitude, and thus such activities are not games.

As Thomas Malaby (2007) points out, we cannot logically use *play* to refer to both a mode of human experience and a form of activity. In other words, we cannot say that when we engage with a game we are entering a particular experiential mode (the lusory attitude, for example) deter-mined by the very act of engaging with the game. As T. L. Taylor (2006) argues, these forms of experientially deterministic arguments oversimplify the complexity of game engagement:

While the notion of a magic circle can be a powerful tool for understanding some aspects of gaming, the language can hide (and even mystify) the much messier relationship that exists between spheres—especially in the realm of MMOGs. ... It often sounds as if for play to have any authenticity, meaning, freedom, or pleasure, it must be cordoned off from real life. In this regard, MMOG (and more generally, game) studies has much to learn from past scholarship. Thinking of either game or nongame-space as contained misses the flexibility of both. (152)

The objection to the magic circle as a form of experiential bracketing has been particularly strong from researchers conducting qualitative studies with players. Ethnographic work by Taylor (2006), Malaby (2007), Copier (2007), and Pargman and Jakobsson (2006) indicates that such a separation is not found in the situated study of gamers:

Problems with using the concept of the magic circle as an analytical tool have made themselves known now and again. These problems become especially clear when the researcher in question has actual empirical material at hand that he or she without much success tries to understand by applying the dominant paradigm of the separateness of play. (Pargman and Jakobsson, 2006, 18)

My own research findings are in line with this view. In analyzing the research data, I found no indication that players enter into an experiential mode that is specific to games. The dimensions of the player involvement model give a more thorough and analytically productive description of game experience without loading it with a priori, prescriptive assumptions about its nature.

Fun?

Huizinga's conception of play as a bounded, ideal space gives rise to another problematic notion that has been inherited by game researchers: an equivalence between gameplay and fun. Indeed, he argues that fun defines the *essence* of play: "Now this last-named element, the *fun* of playing, resists all analysis, all logical interpretation. As a concept, it cannot be reduced to any other mental category. ... Nevertheless it is precisely this fun-element that characterizes the essence of play" (Huizinga, 1955, 3). This seems like an obvious assertion to make: people play games because they enjoy them (Crawford, 2003; Koster, 2005; Salen and Zimmerman, 2003). After all, if games were not fun, why would people play them? There are two problems with this line of reasoning.

First, as Malaby (2007) argues, associating games with fun imbues them with a normative status that ignores the complex and varied experiential states that make them engaging. This does not mean that games are not fun; rather, fun is not an inherent characteristic of games, as has been most generally taken to be the case. As the recent work of Dibbell (2006), Malaby (2007), and Taylor (2006) has shown, contemporary developments in on-line gaming are emphasizing the problematic nature of this assumption. Fun does not denote a specific experiential phenomenon, but spans a whole series of emotional states that vary according to context and individual. As Taylor (2006) states, pinning motivation for game playing on the notion of fun risks missing important aspects of the game experience:

The notion that people play differently, and that the subjective experience of play varies, is central to an argument that would suggest there is no single definitive way of enjoying a game or of talking about what constitutes "fun." We need expansive definitions of play to account for the variety of participants' pleasurable labor and activity. Those definitions must encompass both casual and hard-core gamers. Suggesting that games are always simply about "fun" (and then endlessly trying to design that fun) is likely to gloss over more analytically productive psychological, social, and structural components of games. (70)

Taylor emphasizes the inclusion of labor within the gaming activity, disrupting the common opposition between games and work. As Steinkuehler (2005), Taylor (2006), and Yee (2006a, 2006b) point out, MMOG players often spend extended periods of time engaging willingly in activities which even the players themselves view as tedious or laborious.

A second problem with the notion of fun is that it is too vague an experiential category to be of analytical use. *Fun* merely implies a clustering of positive emotions surrounding an activity; it does not describe what those emotions are or where they derive from. The concept is as unhelpful to the designer as it is to the analyst. It is more productive to focus on a notion of engagement or involvement, which has less normative implications than the notion of fun while allowing further subdivision into constituent elements in a way that fun cannot. Though it is a catchall term that is perfectly adequate for popular parlance, *fun* lacks the qualities of an analytically productive term.

Huizinga's (1955) claim that fun cannot be further broken down analytically is merely an acknowledgment of the vague function it fulfills in language. Huizinga's error is to argue that this vague experiential form is the essence of play, transferring the fuzziness of the concept of fun to the definition of play. The claim of this book is that the experience of gameplay can, in fact, be further analyzed, but this is only possible if we avoid labeling the experience with a priori concepts such as fun or the magic circle. The coming chapters aim to do just that: give a detailed description of the various dimensions of involvement that are present both surrounding and during the moment of gameplay.

4 Kinesthetic Involvement

Macro Phase: Agency

Unlike non-ergodic media, the core experience of digital gameplay requires active engagement and input from players. Players do not merely consume a preestablished piece of media, but instead are active participants in the creation of their experience through interaction with the underlying code during gameplay. The fact that players influence, to varying degrees, what happens in a game environment creates the potential for meaningful exertion of agency. In this section, we will discuss agency in games at the general, or macro, level. Later, we will deal more specifically with avatar and miniature control during players' moment-to-moment interaction.

In its most basic manifestation, agency in virtual environments is the ability to perform actions that affect the game world and its inhabitants. Theorists such as Janet Murray (1998) and Gareth Schott (2006) have viewed agency as the prime contributor to engagement in games. Schott states that "it is the subjective experience of agency that players seem to desire from their engagement with gameplay: they need to feel they have exerted power or control over events" (134). As the player involvement model demonstrates, we should be careful in isolating one form of involvement to the exclusion of others. Nevertheless, it is hard to downplay the importance of agency in a consideration of involvement.

Murray (1998) defines agency as "the satisfying power to take meaningful action and see the results of our decisions and choices" (126). Although knowing that what occurred is a result of intentional action is important, the effects of those intentions are not always as satisfying as Murray claims. Let us take a game-related situation as an example.

In an on-line game of *Counter-Strike: Source* (Valve Software, 2004), I am covering a corridor that leads to a strategic position. An enemy peaks out and ducks back behind cover. I pull out a grenade and lob it into the room he is in. As I throw the grenade, my remaining two teammates rush into the room from a back door that is not visible to me. I run into the room and find the corpses of my two teammates and the enemy. The rest of the opposing team overwhelms me and my team loses the round. The outcome of my action is neither satisfying nor intended, yet it was my exertion of agency that eliminated my teammates and resulted in the loss for our team.

Michael Mateas and Andrew Stern (2005) similarly consider agency as a combination of acting and interpreting responses to those actions. They also claim that a requirement for agency is the existence of perceivable effects upon the world: "A player has agency when she can form intentions with respect to the experience, take action with respect to those intentions, and interpret responses in terms of the action and intentions; i.e., when she has actual, perceptible effects on the virtual world" (para 11).

Mateas and Stern differentiate between local and global agency. Local agency refers to the player's ability to see the immediate reactions to her interaction, while global agency refers to the knowledge of longer-term consequences of a causal chain of events. In my *Counter-Strike* example above, if I had lobbed the grenade into a room and then run off in a different direction, I would still be exerting agency, without necessarily seeing or interpreting the consequences of my actions. This move is often used in *Counter-Strike* as a delaying tactic, stalling pursuers while one relocates to another area or attempts to outflank opponents one might know, or guess, are in the room. My ability to throw the grenade and employ such a tactic in the first place is an expression of agency. My knowledge of the outcome or lack thereof is a question not of agency but of a causal chain of events linked to its exertion. Similarly, I can exert agency in an MMOG that will have effects I may not witness. I could plant a number of mines around a facility in *Planetside* (Sony Online Entertainment, 2003) which eventually could play a decisive role in its defense without my being present at the conflict itself. It is my ability to lay the mines in the first place that expresses my agency in the world, not their eventual usage in action. In this book, I will opt for a more comprehensive conception of agency that is not dependent on knowledge and interpretation of consequences and related satisfaction. In his discussion of agency, Anthony Giddens (1984) emphasizes that:

Agency refers not to the intentions people have in doing things but to their capability of doing those things in the first place (which is why agency implies power: cf. *Oxford English Dictionary* definition of an agent, as "one who exerts power or produces an effect"). Agency concerns events of which an individual is the perpetrator in the sense that the individual could, at any phase in a given sequence of conduct, have acted differently. Whatever happened would not have happened if that individual had not intervened. (9)

Although Giddens views agency as purposive, he stresses the possibility that consequences can be both unknown and unintended. Conceptualizing intended consequences as part of agency diminishes the role of contingency in both everyday life and game environments. The contrivance of contingency is, as Thomas Malaby (2007) has rightly argued, one of the factors that make games appealing for players who specifically value variability and contingency in their play: "It's a huge game. There are so many variants and so many different ways of playing it that I can imagine I'll be playing it for a long time" (Evita, *Planetside*).[1]

Although both games and everyday life contain elements of contingency, the two domains are differentiated by the nature of the rules by which they operate. The rules we encounter in everyday social life are "intended to reduce unpredictability across cases" (8). Game rules, on the other hand, are intended to fulfill a different role: "they are about contriving and calibrating multiple contingencies to produce a mix of predictable and unpredictable outcomes (which are then interpreted)" (9). Malaby formulates games as processes that create carefully designed, unpredictable circumstances that have meaningful, culturally shared, and open-ended interpretations. Therefore, both the game practice and the meaning it generates are subject to change. Players are well aware of the importance of these possibilities in gameplay: "That's why games today are so much better, I think. They give you much more freedom to act and freedom is very important. You can try different things in a game and they actually work in some sensible way. It provokes a creativity, you can feel really involved and [are] doing things" (Evita, *Planetside*).

Without an element of contingency, games stop being interesting, if they can be said to be games at all, as is the case when we learn to play tic-tac-toe "too well." Once all possible permutations of our actions are known, those actions diminish in value. Agency, as it is being viewed here, operates at a more fundamental level. The satisfaction that arises from transforming the contingent into the known or achieved is a consequence of exerting agency. This sense of satisfaction becomes even stronger in the case of MMOGs. One of the most appealing aspects of MMOGs is the potential for actions to have persistent effects experienced by other players that often go beyond those intended. The unintended and unpredictable consequences of one's actions are precisely what can make the exertion of agency in games so meaningful and compelling.

Another aspect of the relationship between agency and contingency is the possibility of failure:

Here the issue is the execution of an action by a participant, an action that may succeed or fail. This kind of unpredictability plays a significant role in athletic contests, but also is the core of many action-oriented computer games. Games call upon you to perform, to accomplish the actions that give you the best opportunity to succeed in the game. At times this performance is embodied and rapid, such as in first-person shooter (FPS) games; at other times, it is simply about not making errors in following game procedures, such as in counting the proper spaces in Monopoly. In a way, all of our actions in games, as in life, are performative in this sense; they run the risk of failure. (Malaby 2007, 16)

The compelling nature of a multiplayer FPS game like *Counter-Strike: Source* (Valve Software, 2004), for example, lies in honing one's in-game abilities: reflexes, hand-eye coordination, and so on. The measure of one's ability is not determined by a set of difficulty levels, but by the abilities of other players. Multiplayer games, particularly competitive ones, increase the social contingency of the game by introducing into the game experience the actions and abilities of other players. This element of shared involvement will be discussed at length in chapter 6.

Different games place varying demands on the player's ability, at times even modulating actions according to the attributes of the avatar. While in *Counter-Strike: Source* the avatar does not affect the player's ability in any way, in *The Elder Scrolls IV: Oblivion* (Bethesda Softworks, 2006) the player's success in fighting is modified considerably by the numerical attributes of his character and equipment. For example, a character with a high sword-wielding skill inflicts more damage per blow than one with a lower skill. Time invested in *Oblivion* has the potential to develop both the player's skill and the character's numerically expressed attributes. The improved attributes or newly gained skills enhance the capabilities of the player, which can result in a heightened sense of agency. The satisfaction derived from this increased sense of agency is not limited to the game world but inevitably extends to the player's sense of personal achievement in general.

In *World of Warcraft* (Blizzard Entertainment, 2004), my skill as a player will take me only so far. Fighting against a player or creature twenty levels higher than me is going to result in certain death for my avatar, no matter how skilled I am as a player. The best I can hope for is to escape the encounter. Similarly, in a duel between two players of equal levels, the player equipped with more powerful weapons and armor will have an advantage over the player with standard equipment, to the point that the relative skill of the players becomes insignificant. *Planetside*, on the other hand, places more emphasis on the dexterity of the player, making it perfectly possible for a level-one character controlled by a skillful player to eliminate a character at level twenty-five (the maximum level in *Planetside*).

If a game rewards numerically quantifiable achievements, motivation and effort will tend to be directed toward increasing these numbers and thus toward one's standing vis-à-vis other agents in the game world, whether human or not: "I just watch the numbers go up ... more levels, more gold, more dps. All I want is more and more stuff ... [and] bigger numbers" (Oriel, *World of Warcraft*). On the other hand, if the game rewards personal skill more than the length of time invested in increasing one's numbers, the motivational drive to hone one's abilities will tend to be stronger.[2] As we will discuss at greater length in chapter 7, both these forms of agency are emphasized when the game world is shared with other human players.

Micro Phase: In-Game Control

Having discussed the general concept of agency in the context of digital games, we can now turn our attention to the specific exertion of that agency in the instance of gameplay, focusing on movement and control. At this point we will need to make a distinction between two types of entities controlled during gameplay: miniatures and the avatar.

The player's presence in many game environments can be fixed to a single entity, known as the avatar. Some games include multiple characters the player can control at different stages in the game, but their control is always associated with a single entity at each moment. Avatars can be interacted with from a third-person perspective, which gives the player some sense of distance, or from the first-person perspective, which gives the player a view of the game world through the avatar's eyes.

In other games, players manipulate a number of entities, or miniatures, sometimes simultaneously and at other times individually. The term *miniatures* refers to entities that players can fully or partially control, but which do not, in themselves, represent the player. Examples of miniatures would be units of cavalry in *Medieval II: Total War* (Creative Assembly, 2006), the workers in *Age of Empires* (Ensemble Studios, 1997), or the falling blocks in *Tetris* (Pajitnov, 1985). I am using the term *miniatures* to account for the perspective this form of control encourages in such games. The world presented to the player is a miniature one, with the player occupying the position of an external, omniscient controller. Avatar control games, on the other hand, beckon the player to inhabit their worlds by anchoring her in the figure of the avatar, which is the primary, if not only, locus of agency within the game environment. This form of control is not anchored to one particular entity but instead embraces the whole environment, as in *Tetris*, or takes on a roving point of view, as in the case of real-time strategy (RTS) games like *Warhammer 40,000: Dawn of War* (Relic Entertainment, 2004).

Modes of Control in Game Environments

Miniatures can be controlled individually or simultaneously. In the case of *Tetris*, for example, the falling blocks are manipulated one at a time. Consequently, the player's attention is focused on rotating and speeding the descent of the falling block (kinesthetic involvement) to the most efficient location, while keeping an eye on the upcoming block (ludic involvement) in order to clear layers of blocks and control the game space (spatial involvement). In *Empire: Total War* (Creative Assembly, 2009), on the other hand, the player controls an entire army, comprising up to twenty units consisting of multiple miniatures each. A crucial part of the game is the simultaneous control of the army to maintain battle lines, while performing strategic tasks to place the player in a more tactically viable position.

Figure 4.1
Commanding large armies in *Empire: Total War* (Creative Assembly, 2009).

The army can be given orders as a whole, in player-assigned groups, or as individual units.

Controlling a single avatar creates a more direct link between the player and the in-game entity. Pressing a key or waving a Wii controller will result in an immediate reaction from the avatar. The relationship between player action and avatar response is thus closely aligned and attention can be devoted to controlling that single entity rather than keeping in mind the locations of multiple miniatures under one's control. As one participant in my research stated, "[I prefer] first person. In [the] first-person view you are usually more in control of the character. You can control almost every aspect of the movement, [such as] aiming, dodging" (Kestra, *Planetside*).

The avatar can take any representational form, from a humanoid figure to a rally car. The most intimate link between a player and even the most unlikely-looking avatar is movement. Even if the player's actions are represented by a vertical bar of white pixels, as is the case in the classic *Pong* (Atari, 1972), the fact that it represents the locus of the player's exertion of agency within the game environment gives a somewhat intimate nature to the relationship it shares with the player. The link between player and game is created through the kinesthetic relationship between player and avatar. As discussed earlier, the ability to exert agency in the game environment is a necessary component of digital games.

Some games require the player to oscillate between controlling multiple miniatures and a single avatar. In *FIFA 2009* (EA Sports, 2008), for example, you control the whole team, one player at a time. Your attention is therefore split between controlling the current player and monitoring the location of other players on the field to whom you can pass the ball. These other players are also under your control, through decisions about formation and other settings. *FIFA 2009* further includes a function called a *player run* in which, at the press of a button, an inactive player will look for an opening in the opposing team's defense so as to receive a through-pass from the actively controlled player. When you do not possess the ball, you can also control multiple players at the same time. For example, if you keep the tackle button pressed, the active player will rush the opposing player who's in possession of the ball, while another player can also assist

Figure 4.2
Controlling two players in *FIFA 2009* (EA Sports, 2008).

in the defensive movement if you press a further button. In effect, this means that you can control the active player to cover or intercept passes, while directing a second player to perform the actual tackle. Thus, even though, at first glance, the basic form of control available is similar to that of traditional avatar-based games, your attention is often also directed to the movement and positioning of other players.

Modes of Game Control

The forms of input currently found in digital games range from the symbolic to the mimetic. It might be useful here to present a descriptive scale of forms of control currently present within games. At one extreme we have the *symbolic control* of keys, controller buttons, and thumb sticks. In this case, there is no direct, mimetic relationship between the actual movement performed by the player and the corresponding movement executed by the avatar. On the opposite end of the continuum of control, we have what can be called *symbiotic control,* in which the player's actions are mapped onto the avatar and have a close relationship with the actions of the avatar in the game environment, perhaps with some modification or exaggeration and inevitably with designed restrictions. The best example of such a control scheme currently available to the gaming market is Kinect (first known as Project Natal). The move toward a more direct symbiosis of avatar and player movement that proved to be the source of success for the Wii console is the main selling point for Kinect, the new control paradigm introduced for the Xbox 360. Kinect utilizes a camera attached to the TV that maps player movement directly onto the avatar; it also recognizes the player's facial features, voice, and distance from the TV screen. If, for example, you are playing a martial arts game, you would utilize your entire body to punch, to kick, and to dodge oncoming blows. In effect, the aim of Kinect is to eliminate controllers altogether. This mapping of the player's physical movements onto the avatar creates a form of control that is substantially different from the pressing of a mouse button or key.

A milder version of symbiotic control is *mimetic control,* a partial mapping of the player's movements onto the avatar. Examples of mimetic control include Wii titles like *Wii Sports* (Nintendo EAD, 2006) and *Grand Slam Tennis* (EA Canada, 2009) in which the player swings the Wii remote in a simulation of an actual tennis stroke. When players swing the Wii-mote, their on-screen avatars swing a tennis racket. This coupling of player and avatar action is the key attractor to the console. In this sense, the Wii expands the kinesthetic dimension to account more meaningfully for both the on-screen performance and the player's bodily activity. Another form of mimetic control occurs when the controller itself replicates part of a machine, tool, or vehicle. Examples of this form of mimetic control are steering wheels for car-racing games and light guns in rail shooters, often found in arcades, such as *The House of the Dead* (Wow Entertainment, 1996).

We can still discuss the dexterity of moving fingers in perfect timing on the *W*, *A*, *S*, and *D* keys in an FPS game, but the attention directed to this form of control is not an involving aspect of the game in itself. If anything, attention tends to be directed to key presses only when a player is not sure which key corresponds to a certain action and needs to consult a manual or Help screen, or learns by trial and error. The difference between mouse, keyboard, or joypad control and the Wii-mote and Kinect interface is that the first three do not share a mimetic relationship with the on-screen action, while the last two involve the reproduction of the players' actions on-screen. The focus on the Wii remote swing is often seen as a pleasurable experience in its own right. The Wii remote, dance mat, or guitar controller can also engage players by turning them into performers for a nongaming audience in the living room. This does not necessarily mean that the players will be more or less engaged, but that the social context within which the game occurs brings an additional dimension to their performance.

Figure 4.3
Forms of control in digital games.

As games develop technologically, we are seeing a shift from symbolic to symbiotic control. Importantly, the further one moves toward the symbiotic end of the continuum, the more appealing the form of control tends to be for casual gamers and nongamers. This is evident both in the forms of games which are developed for platforms like the Wii and the demographic Nintendo's marketing campaigns are clearly aimed at. With the release of the Wii, Nintendo appealed to players of both genders and all generations, particularly the older demographic, which was largely an untapped market in the games industry. Kinect's early marketing videos also emphasize that it is a platform for the whole family, furthering Microsoft's advertised plan to take over the living room with a console that is not solely a gaming machine, but also a media and social hub.

Games often combine control modes, particularly the mimetic and the symbolic. The boxing game in *Wii Sports* (Nintendo EAD, 2006), for example, employs a wholly mimetic control scheme, while the baseball game combines both mimetic and symbolic control when players pitch the ball by making a throwing motion, but press a button to alter the pitch type. A number of other Wii games, such as *No More Heroes* (Grasshopper Manufacture, 2008), utilize the thumb stick on the nunchuk controller to direct the avatar through the environment, while actions are performed by combining button presses with swings of the Wii-mote. This hybrid form of control is necessary in cases where the actions that take place in the game environment cannot be mapped onto the controls in a mimetic fashion. Spatial navigation in a game environment is one example of a control function that is difficult to recreate mimetically.

Kinesthetic Pleasure

The laser cannon representing the player in *Space Invaders* (Taito, 1978) is able to move left or right and can shoot vertically, but has no further potential for motion. *Pacman* (Namco, 1980) moves in four directions on a two-dimensional plane. On the other hand, the title character of *Max Payne* (Rockstar Toronto, 2001), can walk, crouch, sprint, and jump, all while aiming at and shooting his assailants. Further, Max can enter "bullet time," in which the action is slowed down, giving the player more time to execute complex maneuvers akin to the slow-motion acrobatics featured in *The Matrix* (dir. Wachowski and Wachowski, 1999). Fantasies of moving with Neo-like speed are thus partly realized by games such as *Max Payne*, *Enter the Matrix* (Shiny Entertainment, 2003), and *F.E.A.R.* (Monolith, 2005), for, above all, the facility to employ bullet time is an invitation to reproduce on the screen the internal imagery inspired by other media texts that have popularized the concept. Digital games give the player the ability to produce those spectacular moves, and thus also provide them with the satisfaction derived from the performance as well as the cinematic visuals: "[I] love dropping from the air, don't know why, but it's a cool feeling. My favorite part of *Matrix* (Online) was doing those mega hyperjumps from one building to another. You have to calculate where to land from before and there's that excitement of barely making it … and then falling, or not, just like in the first *Matrix* [movie]" (Baal, *Planetside*).

In the case of first-person perspective games, in which the avatar is not visible on the screen, the player's hand and finger actions translate into perceived movement in the world, rather than seeming to be the movement of a manipulated object or character. Thumbing the space key and pressing *W* in an FPS is most often interpreted by the player as a forward

Figure 4.4
Max Payne (Rockstar Toronto, 2001)—life in bullet time.

jump they have made, rather than as the forward jump of a character they control.

Players experience a double awareness in third-person perspective games. First, they are aware of the surrounding environment as portrayed by the camera. Second, they are also aware of the space as it relates to the avatar. In games that allow for switching between points of view, the player must remember that her ability to see around a corner does not necessarily mean that the avatar shares this line of sight. Thus, when switching back to first person, the player needs to move in order to look around the corner and shoot, for example. As one participant explained, "It depends on the game and what you are trying to do. Like for *Planetside*, third person helps you to get a good look at what is around you, but you have very little aim control, while first-person view gives you that soldier's-eye view so you can aim and you get a sense of actually being there ... controlling that character" (BunkerBoy86, *Planetside*).

Movement is therefore an essential part of gameplay: It is the key ingredient that allows players to act upon the environment and is thus a necessary condition for the sense of agency that is a crucial factor in the game experience. It is not only a central component of the ludic aspects of gameplay but is also an enjoyable part of the experience in its own right, particularly when the controls have been mastered and a fluent engagement with the environment is enabled. Part of this pleasure is the ability to simulate experiences that are not possible in the physical world, as the research participants described so vividly:

Nothing I like more than hopping from building to building on my flight max. Now they got nerfed,[3] so that sucks, but the cool thing is being able to do those big jumps, seeing buildings and people becoming smaller and then bigger again when you land on them especially if you're blazing your chain gun as you drop. (Baal, *Planetside*)

Participants often described the sensation of movement as if they were experiencing it directly. They mentioned making an additional imaginative effort to bring about or enhance these sensations. One participant, Sunniva, used the term *flow* to describe how the sensation of movement inside the game world holds her attention and creates a sensation of "continuous focus," which she likened to surfing: "[It's] like a wave—a continuous motion that keeps my attention on the game ... it's almost kinda peaceful—well, in the sense that it creates a certain steady, continuous focus ...

maybe it's kinda like surfing—hehe—although I've never done that" (Sunniva, *Planetside*).

Another participant, Bladerunner, used the term *Quan* to describe the internalized dimension of kinesthetic involvement. As we discussed in the previous chapter, *internalized* involvement describes a situation in which no conscious attention is being directed toward the dimension in question. The internalization of kinesthetic involvement describes a situation in which the controls are learned to such a degree that the on-screen movement of avatars and miniatures feels unmediated: "The Quan can only take place when you feel your actions on the screen are taking place not because you're pushing keys on your keyboard and moving your mouse, but [because] your mind is willing those actions" (Bladerunner, *Planetside*).

Participants described the intensity of engagement fostered by the disappearance of mediation in relation to movement, citing it as one of the major sources of gaming pleasure: "The feeling like you're really holding that gun. And that's what rocks my world" (Baal, *Planetside*). Sue Morris (2002) emphasizes that this internalization of controls is a necessary step toward mastery in FPS games:

The skilled player's hand movements controlling the keyboard and mouse also become unconscious to a degree. Some manoeuvres require a complex sequence of keystrokes and mouse movements to be performed. These are analysed, designed, memorised and practiced, but beyond this they are internalised, to the point at which individual moves are not carried out consciously and individually, but are absorbed into the player's style of playing. ... To achieve mastery of the game, conscious playing is not enough, players must be able to "think with their fingers" to the point at which the player feels like an extension of the game or the game feels like and extension of the player. (85)

The fluid continuity of control between player and avatar creates the potential to awaken kinesthetic sensations in players that make the game-playing experience pleasurable in its own right. One reason that running, leaping, or flying through a game environment can be so appealing when mastery of the controls has been achieved is that players can then physically and cognitively interpret those actions as *their own* running, leaping, and flying, rather than as those of an external agent: "I am now flying an aircraft and there is a dude down there shooting a rocket at me. ... At those moments I don't say it's my character being under fire or shot, I always say's it's me" (Evita, *Planetside*).

In this respect not all games are created equal, as these sensations are related to the forms of spatial navigation and movement afforded by the game environment. *Mirror's Edge* (DICE, 2008), for example, is essentially a game about creating a continuity of movement through an obstacle-filled space. The game rewards the perfect timing of a player's movements by giving him faster sprinting, the ability to jump farther distances, and greater momentum. Like the urban sport of parkour, from which the game draws its inspiration, *Mirror's Edge* transforms obstacles in the surrounding space into affordances for fluid and rapid movement.

During her research with local area network (LAN) players, Melanie Swalwell (2008) similarly found that her participants greatly enjoyed the sensation of movement, particularly when they had achieved mastery of the controls. One problem with Swalwell's argument, however, is that she does not distinguish between the onlooker and gamer in her consideration of kinesthetic engagement. She cites Van Rooyen-McLeay (1985) who, in her master's thesis, observed physical excitation during arcade gameplay:

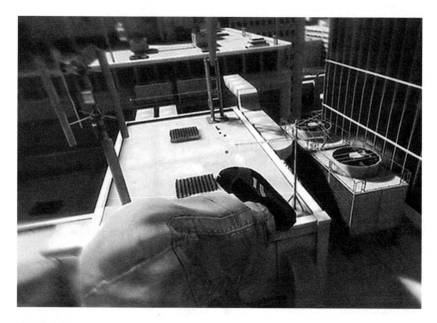

Figure 4.5
Movement in *Mirror's Edge* (DICE, 2008).

This player's actions clearly represented what I had seen so many times before. His hands furiously beat the buttons on the machine and his legs uncontrollably "danced" beneath him, as he attempted to come to grips with the excitement of the game ... occasionally, an onlooker would let out a wail, or his or her legs would twitch with excitement. (Van Rooyen-McLeay, cited in Swalwell, 2008, 74)

Swalwell emphasizes the stimulation derived from the game across players and viewers. Van Rooyen-McLeay's observation of excited reactions to arcade gameplay by both the players and onlookers does not necessarily mean, however, that the nature of engagement is equivalent. The onlookers are excited about both the occurrences on the game screen and the visible excitation of the players themselves. The players, on the other hand, are also, to some degree, aware that they have an audience of onlookers which in turn affects their level of excitement. Importantly, though, the onlookers are not part of the circuit created between game and player, nor do they have a sense of agency in the game environment. The failures and accomplishments of control are not theirs to bear or celebrate. Swalwell does not account for these differences in her paper; in fact, at times she implies that the purely perceptual experience of a medium's representational layer is the same as the combination of this aspect of games with their ergodicity.

The focus on control and affordances for movement designed into game environments means that understanding an experience such as kinesthetic involvement requires a simultaneous consideration of the visual, audio, and haptic output along with the disposition to act on the part of the player. The disposition to act, and not solely the instance of input, is critical because, as explained in chapter 3, game involvement is not limited to the flicking of a thumb stick or clicking of a key, but also concerns the active intention to perform the action at the right time. In chapter 3, we saw the example of a sniper in *Red Orchestra* (Tripwire Interactive, 2006) covering a vacant street. In the seconds or minutes of waiting, the sniper might not be particularly active, yet waiting for a target leaves her tense with anticipation to move the crosshairs, lead the target, hold her breath, and press the trigger button at the right moment. The expectation of performance can be just as engaging as the performance itself. Onlookers do not experience this level of anticipation and agency. They might attempt to participate by calling out suggestions to players, but their kinesthetic experience is confined to the mimetic, while players combine the mimetic with the ergodic to arrive at a more intense and qualitatively different kinesthetic experience not available to the person watching a game being played by others or a film.

Kinesthetic involvement relates to all modes of avatar and miniature control in virtual environments, ranging from the process of learning the controls to the fluency of internalized movement. This dimension of involvement requires more conscious effort when the controls catch the player's attention, either because they are not fully mastered or because a situation demands a complex sequence of actions that are challenging to the player. Players of varying skills and preferences will be more or less involved by different kinds of movement. Some love going as fast as possible down a racing track. Others lose themselves in executing multiple barrel rolls in World War I biplanes. Some become deeply engrossed in coordinating their actions with those of other players in multiplayer FPS games, while others become most involved when sneaking patiently and silently through an area infested by enemies who are unaware of their presence. Still others enjoy a leisurely ride on their mount in *World of Warcraft* (Blizzard Entertainment, 2004), taking in the beautifully designed landscape. The important thing to consider is that, in all cases, movement is a crucial part of the gaming experience. The freedom of action allowed as well as the learning curve of the controls concerned will have a major influence on the players' ergodic involvement in the game environment, as most other aspects of involvement in games are dependent on developing at least a basic fluency of movement in the environment.

5 Spatial Involvement

Macro Phase: Exploration

The desire to explore new lands has been an inherent part of human nature since the beginning of our species (Lewis-Williams, 1998, 2002; Evans, 2001). The places we yearn for most are those that are different from our everyday surroundings, especially as promoted and popularized by the media. A resident of an urbanized country with scant greenery might yearn for the natural beauty of New Zealand, particularly after its association in the popular imagination with Tolkien's Middle-earth in Peter Jackson's film version of *Lord of the Rings* (2001). New Zealanders, on the other hand, might romanticize the old buildings and cobbled streets of Europe, longing for the ambiance that centuries-old architecture tends to create and which is not found in such a young country as New Zealand. Despite this desire to explore other places, many people are daunted by the idea of leaving the comfort of familiar surroundings to actually visit strange lands. Aside from this, there are always constraints related to job and family commitments, financial costs, and other lifestyle limitations that get in the way of such travels.

One way of appeasing this penchant for exploration is through visual media. Digital games and virtual worlds are particularly adept at facilitating spatial exploration that enables players not only to project their imagination into the represented landscapes but also to traverse them. As the geographer Yi-Fu Tuan (1998) argues, in an age of urbanized living, the rural and pastoral have a tendency to gain in attraction. This romanticized view of nature might explain the attraction of fantasy settings such as Tolkien's Middle-earth and, consequently, the dominant presence of the fantasy genre in MMOGs. With the considerable advances in graphics and audio technology that have taken place, it is now possible to create vast virtual worlds in which the desire for active engagement with unfamiliar situations and settings can be satisfied. As one participant in my research interviews explained: "I like foot travel for the sightseeing aspects: exploring new areas with new environments. ... I have to say that I really enjoy finding new areas, the scenery. I've actually experienced 'tourist' moments in some games. I've found a spot with such a great view I wanted to take a picture" (Rheric, *World of Warcraft*).

Contemporary digital games offer players geographical expanses to inhabit, interact with, and explore. Certain game worlds invite players into picturesque landscapes which give the impression that they extend as far as the eye can see. This gives players a scope for exploration and discovery that is made all the more appealing if their lifestyle prohibits travel to exotic new lands:

There is so much [virtual] world to run around in that every day is filled with new discoveries. The enormous scope of the worlds draws in players who feel stuck in the shrinking, overcrowded, polluted real world. In many games, it's common, for example, for players to pause occasionally on a snow-capped mountain peak just to look out over the forests that stretch for miles in every direction down to the palm-fringed tropical beaches, parched deserts and vast steppes. (Kelly, 2004, 63)

Richard Kelly's sentiments on exploration resonate with those of research participants who were attracted to the scope for exploration of MMOGs like *World of Warcraft* (Blizzard Entertainment, 2004). A number also felt a sense of wonder upon reaching areas affording attractive vistas which one participant, Rheric, calls a "tourist moment." Another participant similarly associated the beauty of virtual landscapes with tourism: "I like *World of Warcraft* 'cause it's pretty. You come over a hill and, bang, you're struck by the scenery like you're on holiday somewhere exotic" (Baal, *World of Warcraft*).

What is attractive is not only the beauty of the landscape but the element of pleasant surprise at making the discovery. There is an important difference to be appreciated between ergodic, simulated landscapes and non-ergodic representation of landscapes. Although one can imagine roaming around the represented space described in a piece of literature, traversal is limited to mental imagery. To move from one point to another in a game world, players must literally navigate their way, not merely imagine it.

In avatar control games, players have an embodied presence in the game world that has a front, back, left, and right. Much as in the real world, there is an orientation of a body in space with a sense of a future ahead and a past lying behind (Tuan, 2007). This presence can be more loosely organized in a virtual environment than in the actual world, since in some cases players can opt to view the world in the first person or to zoom out to a third-person perspective. Nevertheless, the avatar still has an orientation to the world around her through which the player's agency is channeled. When a player plots a route through a geographical expanse and then navigates it, it is more likely that she will feel a sense of habitation within the game environment. There is the added satisfaction of having expended effort to reach a particular destination, especially when reaching this goal is challenging:

I have played a game where if you wanted to get to a great place it entailed running on foot for a half hour. It made me feel frustrated on one level and really happy once I got there, an achievement of sorts. I would like to see more land masses to explore: Mountains that are traversable only on foot or require climbing gear. Risk. The sort of things that make travel in real life more engaging. ... The islands and areas on the map that aren't currently open to the public (it's the mystery of it). I have died trying to swim to them. (Ananke, *World of Warcraft*)

Lost in the Game World

A journey becomes more rewarding when the player does not command an all-knowing perspective of the space around her and runs the risk of getting lost. The possibility of getting lost only exists in a virtual environment whose structure yields a degree of spatial contingency. If carefully contrived by designers, this spatial contingency can yield a sense that the simulated space extends beyond the confines of what can be immediately seen. This, as Lisbeth Klastrup (2004) describes in her doctoral thesis on the subject, is a defining element of virtual "worldness." If the virtual world is marked by paths, those paths should have an expanse of surrounding environment to retain the sense of being a landscape capable of exploration. The moment players realize that there is no opportunity to become lost, the scope for exploration is severely diminished and the environment is perceived for what it is: a multicursal labyrinth (that is, one with branches and dead ends).

As Espen Aarseth (2005) has observed, labyrinthine spatial structures are very common in games. Once unraveled, the labyrinth contains a linear progression which makes it easier for the designer to control the pace at which players encounter embedded events, characters, objects, quests, information, and segments of scripted narrative.[1] The labyrinth constrains players' orientation by offering only forward and backward movement: what has come before and what is to come. It excludes the 360-degree orientation that the sides of a body enable. As Yi-Fu Tuan (2007) describes the experience:

What does it mean to be lost? I follow a path into the forest, stray from the path, and all of a sudden feel completely disoriented. Space is still organized in conformity with the sides of my body. There are the regions to my front and back, to my right and left, but they are not geared to external reference points and hence are quite useless. Front and back regions suddenly feel arbitrary, since I have no better reason to go forward than to go back. Let a flickering light appear behind a distant clump of trees. I remain lost in the sense that I still do not know where I am in the forest, but space has dramatically regained its structure. The flickering light has established a goal. As I move toward that goal, front and back, right and left, have resumed their meaning: I stride forward, am glad to have left dark space behind, and make sure that I do not veer to the right or left. (36)

If we apply this perspective on orientation based on the body to virtual environments, we can understand simulated space not based solely on our avatar's spatial location, but also based on the goal-directed movement which reinforces the tangibility of our avatar by giving it a meaningful front and back, right and left. A goal can be indicated by a quest, set by an in-game agent, or set by the player herself. The possibility of getting lost or not succeeding heightens players' spatial involvement. Besides making the individual journey more engaging, this risk adds a level of excitement at the idea that the future holds the promise of similar discoveries: "The first time I ran from Menethil Harbour to Ironforge as a young Night Elf Hunter was pretty exciting. I almost got eaten by a large crocodile and had to dodge some orcs. The danger involved [was exciting], and exploring new areas for the first time" (Inauro, *World of Warcraft*).

Participants often expressed their desire to explore the unknown, to unravel the mysterious and out-of-bounds. In digital game environments, this desire is manifested not only through experiencing the topography of a landscape, but by pushing the boundaries and if possible subverting what designers intended players to do within the spaces they created. The game world is thus, in the words of Jorge Luis Borges (1972), a "labyrinth devised by men, a labyrinth destined to be deciphered by men" (42). The designed nature of game worlds beckons players to traverse and unravel them, to explore and understand not only their representational domains but also the collective minds of their designers. Thomas Malaby (2007) argues that one of the most compelling aspects of games is their ability to represent the contingent nature of our existence. Compare a maze to the uncharted bush. No matter how confusing the maze is, its traversal promises a solution. The contingency of its traversal is contrived by the designed existence of a correct route and accompanying dead ends. On the other hand, the contingency present in the traversal of uncharted bush is of a different order. It has not been created with a specific solution and thus it cannot be read with such an aim in mind. The design of game worlds thus involves the creation of "complex, implicit, contingent conditions wherein the texture of engaged human experience can happen" (Malaby, 2007, 15). What makes travel in virtual worlds appealing is not only the affective power of their aesthetic beauty, but also the performed practice of exploring their technical and topographical boundaries.[2]

Micro Phase: Navigating Space

In the above discussion, we considered the attractions of spatial exploration and the pleasures of discovering and getting lost in a game environment. The micro phase focuses on three forms of spatial engagement: learning the game environment through navigation, mentally mapping the navigated space, and controlling miniature space through tactical interventions. Before I discuss these three forms of spatial engagement, it will be helpful to set out the broad categories of spatial structures that feature in games.

The concept of *navigable space* will be used throughout the rest of this chapter to refer to the dominant form of space found in game environments, in which players must move and act. Mark Wolf (2009) has defined navigable space as "a space in which way-finding is necessary, a space made of interconnected spatial cells through which the player's avatar moves, a network often organized like a maze. All of the space may be present onscreen at once, as in certain maze games, but typically much of the space resides off-screen, and the accessibility, and even the discovery of off-screen areas, relies on a player's navigational ability" (para 7).

Navigable space is thus necessarily simulated space, as opposed to a space that exists only as a representation. Our concern in this chapter is primarily with navigable, simulated space.

Spatial Structures: Unicursal Corridors or Labyrinths

The opening sequences of titles in the *Medal of Honor* series of World War II FPS games unmistakably plunge the player in the midst of a hectic battle scenario. The first moments of *Medal of Honor: Allied Assault* (2015 Inc., 2002) bear an uncanny resemblance to the beach landing scene in the movie *Saving Private Ryan* (dir. Spielberg, 1998). The game's sequel, *Medal of Honor: Pacific Assault* (Electronic Arts and TKO Software, 2004), is set in Pearl Harbor, minutes before the Japanese assault. The pace of the action is frantic: shells explode all around, peppering the screen with shrapnel and sand. Injured soldiers cry out for assistance and the squad leader's orders are barely audible over the sound of gunfire and bombing. In a short prelude to this scene, the player's avatar is driven around Pearl Harbor by an officer. The path the jeep takes is predetermined and the only thing the player can do is look around the immediate environment. Although the rest of the game gives the player more freedom to move than this initial sequence, traversing the scenarios is a strictly linear affair. Most environments are far smaller than they look. In the first level of *Medal of Honor: Allied Assault*, for example, the beach appears to extend left and right for miles, with cliffs standing as barriers to exploring the area of land beyond them. In effect, however, there is only a single path the player can take through this area. The pressure from bunker fire, strafing planes, and the general chaos instills a sense of urgency in the player, making it easier to conceal the limited traversable space in the scene. As players explore the spatial limits of the game environment, they form a different view of the space represented: a superimposition of the internalized model of the space with that represented on the screen. For example, after trying to approach the German bunkers from the west side of the beach, players of *Medal of Honor: Allied Assault* realize that they are repeatedly killed by artillery fire, no matter how hard they try to stay alive. As a result, they will tend to try another route. The realization that the western flank is impassable, even though it looks traversable, will result in an updated cognitive map of the environment to take account of this invisible wall. The concept of such cognitive mapping was introduced by

Edward Tolman (1948) and can be understood as "the process of encoding, storing, and manipulating experienced and sensed information that can be spatially referenced" (Golledge and Garling, 2004, 502). The cognitive map is thus a mental model that accumulates and organizes spatial information into a coherent, internal image that facilitates both learning and recall. Cognitive maps are accumulated through navigation in the environment as well as more general surveying by using vantage points, maps, and other spatially holistic sources. As Liboriussen (2009) argues, cognitive mapping is also important in virtual-world navigation: "The cognitive map is a mental tool that aids its constructor in navigating the virtual world. At the same time, the cognitive map provides an overall sense of how the virtual world is structured and a sense of connectedness" (193).

Unlike the physical world, where a distant location is reachable only if we find a means of navigating to it, game worlds often present locations that are not, in fact, reachable. In the *Medal of Honor* example above, the incessant barraging of artillery fire on the left flank cannot be turned off or surpassed in any way, and thus its presence is taken into account in the players' cognitive mapping of the space ahead of them. The illusion of an open world is undermined when players realize that the represented world beyond the immediate, corridor-like area does not afford traversal. As Aarseth (2005) argues, it is quite common for game spaces to give the impression of being open when they are in fact "unicursal corridors" (499) or labyrinths.

Figure 5.1
The opening sequence from *Medal of Honor: Allied Assault* (2015 Inc., 2002).

Michael Nitsche (2008) outlines a number of spatial structures in his book *Video Game Spaces: Image, Play, and Structure in 3D Game Worlds*, distinguishing between unicursal labyrinths and an alternate spatial structure he calls "tracks and rails" (172). Tracks and rails are most clearly visible in racing games such as *Gran Turismo* (Polyphony Digital, 1998), but Nitsche also includes shooting games like *Medal of Honor* that guide the player down a single available path. The distinguishing feature between these two, according to Nitsche, is that labyrinths place their restrictions on display. The structures outlined here are based primarily on traversable space, not visible space. This makes it unnecessary to distinguish between tracks and unicursal labyrinths in the way Nitsche does, since both structure *navigable* space along a single path.

Clara Fernandez-Vara (2007) distinguishes between labyrinths and maze structures in games, stating that labyrinths consist of a single winding path through their domains and hence "labyrinths do not present a test to their visitors" (75). Of course, although this might be true of classical labyrinths, game labyrinths always contain some form of adversity. This could take the form of a puzzle the player needs to solve before being allowed to progress or a group of antagonistic agents who are blocking the way. *Half-Life 2* (Valve Software, 2004) incorporates both forms of barrier in an alternating fashion, creating moments of adrenalin-pumping action followed by slower-paced spatial navigation and physics-based puzzles.

As Geoff King and Tanya Krzywinska (2006) point out, the restrictions of movement in labyrinth structures should not be viewed solely in pejorative terms, for the restriction of spatial freedom allows designers to have tighter control on other aspects of involvement. As we will discuss in chapter 7, the scripted narrative written into a game is more easily conveyed to players in more spatially restricted environments. Similarly, emotionally charged situations are easier to orchestrate, since designers can count on players triggering a particular event in a specified location.

Spatial Structures: Multicursal Corridors or Mazes

Mazes, on the other hand, offer multiple routes through their domains, but in games these routes have been predesigned by the level designers. Common features of mazes are their branching paths and dead ends, which encourage players to pay attention to the navigational choices they make and to mentally map the pathways of the environment (Fernandez-

Vara, 2007). Yet this mental mapping is necessary only when the maze is presented to players in first- or third-person perspectives and extends in a space that is not contained by the individual screen, as in *Fable* (Lionhead Studios, 2004). Here again, players are given the illusion of a large open world, but a closer look at the spatial configuration of the game world will immediately identify it as a multicursal corridor or maze structure. Players are channeled through pathways created by steep hills and rock formations that cannot be traversed. The outer edge of the map is surrounded by desert, but if players go too far into it they collapse of heat exhaustion, effectively barring them from exploring its vast space and relegating it to a traversable ring around the hilly interior.

The major difference in spatial structure between *Fable* and omniscient perspective games like *Pacman* (Namco, 1980) is the anchoring of the player's perspective to a location *within* the environment of the former as opposed to the ability to freely roam the landscape as an invisible eye in the sky that the latter allows. Omniscient point-of-view games whose spaces extend beyond the screen involve a certain amount of spatial exploration through scrolling and thus open up the possibility for players to learn the environment's spatial layout, but the form of cognitive mapping of space that occurs in these cases is more readily comparable to getting an indexical sense of the layout of an area from Google Maps than to learning the space through navigation.

Spatial Structures: Rhizomatic Zones

Most game spaces are not spatially contiguous, and so players often teleport from one location to another. In such arrangements, there is often no way to reach the two points without teleporting. In *Fable II* (Lionhead Studios, 2008), for example, players can automatically teleport to certain predefined points within different areas as long as they have already discovered the location in their travels. In *Mass Effect* (Bioware, 2007), on the other hand, the different locations cannot be reached by the continuous traversal of space. The spaces in *Mass Effect* are separate zones, which can be accessed only by teleporting into a designated entry point. We can call such a spatial structure *rhizomatic*, following Gilles Deleuze and Félix Guattari's (1987) elaboration of the concept of the rhizome in *A Thousand Plateaus*. They define a rhizome as a system made up of linked points, without a hierarchy or center, forming a multiplicity with no beginning, center, or end. A rhizomatic system can connect any two points on its structure without the need of approval by any centralized power. It can be accessed via multiple entry points and due to its non-linear arrangement it cannot be broken since if it is "shattered at a given spot, ... it will start up again on one of its old lines, or on new lines" (9).

Even when the game space is represented as being contiguous there are often breaks in the flow of navigation as new zones are loaded into existence. At times these breaks in navigational flow are concealed by channeling players through locations in which the interruptions make logical sense within the fiction of the world. For example, in *Dead Space* (EA Redwood Shores, 2008), play is punctuated by journeys in elevators that connect the various floors of a level. The loading of a new zone is concealed by the elevator journey, during which the player can still perform actions such as checking their inventory, reloading their weapons, and so on.

Spatial Structures: Open Landscapes

The spatial structures which afford players the most freedom for navigation are "open landscape" environments (Aarseth, 2005, 499). An example is *The Elder Scrolls IV: Oblivion* (Bethesda Softworks, 2006), a fantasy RPG which offers an extensive landscape for players to explore. Players of *Oblivion* are not forced to move through a number of limited environments in a linear sequence. MMOGs also tend to offer similarly large open tracts of land to be explored. Any restriction on traveling between various areas tends to be related to the difficulty of overcoming aggressive inhabitants rather than to spatial constraints. Hobbits in *Lord of the Rings Online* (Turbine, 2007), for example, start their adventures in the Shire. They are free to attempt to journey to the city of Rivendell, but until they are sufficiently powerful they are likely to get killed on the way. With assistance from other high-level friends, they may, however, succeed in their journey. The spatial constraints are therefore more related to the game rules than spatial structures.

Within the category of open landscapes, MMOGs deserve particular consideration. Edvin Babic (2007) argues that MMOGs present a departure from earlier spatial design in games that was almost entirely instrumental. This argument is based on his thesis that the design of game spaces has evolved in a similar manner to architectural planning theory, which shifted from an "object-centred and deterministic view on space into a pluralistic and culturally sensitive appreciation of the relation between social practices and material acts" (para 4). Following the German sociologist Martina Löw (2001), Babic calls the first conception "absolute space" (para 6) and the second "relational space" (para 6). Absolute space emphasizes a Euclidian view of space as an empty container that expresses different arrangements of bodies and objects in an objectively measurable manner. Relational space, on the other hand, exists "through the context given by the relations and interactions of the actors and objects within" (para 8). According to this more recent conception of space, discussed by theorists like Henri Lefebvre (1991), Edward Soja (1996), and Yi-Fu Tuan (1974), space is seen as socially constructed rather than formally defined. Babic associates pre-MMOG spaces with the absolute perspective on spatial design. Here, the design of game space was structured to accommodate a specific activity. MMOGs, in contrast, bring out the qualities of relational space by allowing for a transformation of space through persistent social interaction.

Babic's argument can be framed through the dimensions of the player involvement model. In the case of MMOGs, the spatial involvement dimension overlaps more intimately with the shared involvement dimension, while in the case of other games he mentions, such as *Doom* (id Software, 1993), spatial involvement overlaps more frequently with kinesthetic involvement.

Spatial Structures: Arenas

Michael Nitsche (2008) identifies the arena as a particular form of spatial structure that is commonly found in games. Nitsche usefully differentiates the spatial qualities of arenas from those of other spatial structures, noting that rather than providing visual cues for navigation, they instead "provide the canvas for a performance" (183). Arenas are sometimes embedded in larger virtual worlds like the battlegrounds in *World of Warcraft* (Blizzard Entertainment, 2004). Although arenas can have any of the spatial structures outlined above, spaces intended to host multiplayer battles generally have different design criteria than other game spaces. This is because the competitive nature of this subgame within the larger game environment of *World of Warcraft* requires a degree of balance to promote fairness for the different factions involved.

Players tend to develop a more intimate knowledge of such spaces, since such knowledge gives them a considerable edge in the competitive element of the game. These spaces are also more easily learned because they tend to be smaller in size than other spatial structures. In the first-person shooter (FPS) and real-time strategy (RTS) genres, the spaces for competition are called *maps*, while in racing games they are called *tracks*, and so on. Arenas include the simple rectangular shape of a football field, the high-speed tracks of a car race, the maps in FPSs, and the arenas embedded in MMOGs.

Arenas are possibly the spatial structure of which players have the most intimate immediate knowledge. In an open landscape in which one can freely roam, players rarely need to learn the layout of their immediate area to the level of precision that players in arenas tend to develop. In a car-racing game, for instance, it is crucial to have a sense, whether conscious or not, of the velocity one can take a corner at, since the game revolves around maximizing one's speed. Similarly, in a game of *Counter-Strike: Source* (Valve Software, 2004), it helps players greatly to know the layout of the map, the most common hiding spots for their enemies, and the areas that afford a good perspective on key zones of the map while offering protective cover. It also helps to know how movement is related to space in such games. Experienced players of *Counter-Strike: Source* will know how much ground they can cover if they run the shortest route with the lightest weapon (the knife) in their hand as soon as the round begins. Similarly, it is crucial for these players to know where they will meet the opposition if someone on their team adopts the same tactic, known as "rushing" a particular objective.

This intimate knowledge of the map is required in multiplayer competitive games. The tactical elements of the game are heavily influenced by the spatial layout of the map in relation to the capabilities of the player, her avatar's qualities, and the tools she has at her disposal. For example, as we can see from the overhead view displayed in figure 5.2, if I start a round of *Counter-Strike: Source* with a shotgun and I know that the opposing team is armed with rifles and sniper rifles, I will be sure to engage the enemy at close range because their weapons are less effective than mine at that distance. I will therefore move from cover to cover and not expose myself to areas that offer a long corridor of fire, hoping to get the jump on the enemy in a constrained space. Thus, knowing the de_dust2 map and starting out on the terrorist side, I will avoid running down the central corridor route as well as the right-hand route to bombsite A because they both provide long corridors of fire for the counterterrorists. On the other hand, the route to bombsite B and the amount of cover available there will play to the strengths of my weapon.

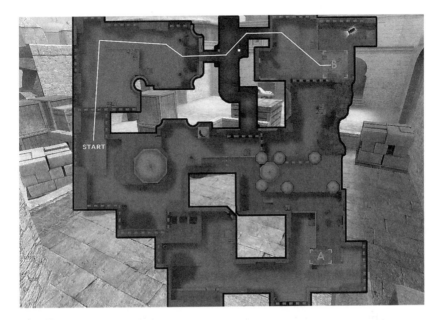

Figure 5.2
de_dust2 B rush in *Counter-Strike: Source* (Valve Software, 2004).

As this example illustrates, knowledge of the spatial features of the map has an important relationship to the rules and physical properties of the game in question. In some FPSs, for example, a single shot to the head is enough to kill the player and eliminate him from the round; such is the case in *Counter-Strike: Source*. Other games, such as *Unreal Tournament* (Epic Games, 1999), require a number of shots for a player to be killed, making it less dangerous to be out in the open. In games like *Unreal Tournament*, in which players can move at relatively fast speeds and can take more punishment, constant movement is more conducive to survival. In contrast, in *Counter-Strike: Source*, it is during bursts of movement between cover that players are most vulnerable. Thus, players of *Counter-Strike: Source* base their movement on the cover afforded by the environment to a greater degree than players of *Unreal Tournament*.

From Space to Place
All the above forms of spatial configurations in games encourage players to internalize their immediate location within the map (*Counter-Strike*), level

(*Medal of Honor*), region (*The Elder Scrolls IV: Oblivion*), or world (*World of Warcraft*). This cognitive mapping of traversable game space works on various scales, ranging from locating oneself in the immediate visible area to identifying one's location vis-à-vis the larger geographic context. An example of the former can be seen through a situation in *Counter-Strike: Source* (Valve Software, 2004). I am playing on the terrorist side on the de_ dust2 map. As soon as the round starts, I run toward bombsite B through an internal area. I go through the doorway and turn left, only to hear the sound of a flash grenade bouncing into the room from the left doorway. The screen goes white and the sound is blocked out by a high-pitched whine. Knowing the map layout fairly well, I press the *D* key to move right and then the *S* and *D* keys simultaneously to make sure I'm moving along the back wall and going down the steps. I then move the mouse left and let go of the *D* key so that I'm facing up the steps when the flash grenade's effect wears off, as in all likelihood the counterterrorist who threw the flash grenade will be running into the corridor looking for a dazed opponent.

On a broader geographical scale (where one exists), spatial involvement relates to one's location within a larger world and the possible methods for traversal. The inferred size of the world gives players a greater sense of grandeur, an indication that beyond those mountains there are further lands to be explored whenever the player feels inclined or able to do so. This creates a sense of place that helps make the world a more believable, habitable place rather than simply a chain of environments linked together, as is the case in games like *Medal of Honor: Pacific Assault* (Electronic Arts and TKO Software, 2004), or a series of disconnected locations, such as the maps in *Counter-Strike* (Valve Software, 1998).

Figure 5.3
Navigating through a space blindly in *Counter-Strike: Source* (Valve Software, 2004).

As cognitive maps of game environments improve, the player's spatial disposition to them shifts from the conceptual to the inhabited. As the lay of the land is memorized, less investment of conscious attention is required to orient oneself in the virtual environment. The process of internalization involved in learning the layout of a map, region, or world leads to a stronger sense of inhabiting the game space; if looked at from a different perspective, it leads to the game environment becoming part of the player's immediate surroundings:

I am a pretty visual person. I drive by landmarks not by road names, so, I can visualize it pretty vividly. Though some of the in-between spaces are hazy cause I fly from Undercity to Tarren Mill now. It doesn't matter if it's the real world or digital, as I travel around, and learn new areas, I naturally seek certain kinds of landmarks to help me keep my bearings. A twist in the road here, a tree there. I find it comforting when I start to get the lay of the land in an MMO. For others, I have no idea. It's not just comfort, but also a bit of pride. I know where I am, I know where I want to go, and I know how to get there. (Rheric, *World of Warcraft*)

Rheric associates exploration and knowledge of the land with a sense of comfort and pride. The shift toward internalized spatial involvement creates a greater sense of habitation and belonging to the region. This internalization process is crucial for the achievement of a sense of comfort and place both in the physical and the digital worlds, as Rheric states. If you move to a new city in a country you've never visited, you first begin by mapping the surrounding area, creating a mental image of where things are. In these initial phases, people tend to feel lost and disconnected from the place. Cognitive maps of the area start being built on the basis of chains of landmarks and recognizable routes one moves through over and over again. As these are learned, a sense of comfort and belonging settles in, creating an attachment between player and game environment. This sense of attachment and familiarity can be seen as the distinguishing factor between perceiving an environment not only as space, but also as place: "When space feels thoroughly familiar to us, it has become place. Kinesthetic and perceptual experience as well as the ability to form concepts are required for the change if the space is large" (Tuan, 2007, 73).

This process can occur in virtual as well as actual environments. In digital worlds, the primacy of visual information and cognitive map formation play an even more crucial part than similar processes in real life, because nonvisual cues like heat, smell, and variety of sound are either absent or restricted in the digital world. Thus, the visual aspect of mapping digital game worlds becomes even more salient because it is the primary means of identifying and distinguishing space. The speed and efficiency of this process of internalization is dependent on the individual but is strongly influenced by the design of areas in the world and how they can be traversed or manipulated. Such design issues become particularly important in large worlds such as Azeroth in *World of Warcraft* (Blizzard Entertainment, 2004) or Norrath in *EverQuest* (Sony Online Entertainment, 1999).

World geographies that have easily distinguishable areas with clearly delineated boundaries make it easier for players to remember and internalize the world layout. The more readily players assemble a cognitive map of the continent, world, or area, the easier it is for them to become, as Rheric points out, more involved in the geographical aspects of the game world. Lynch (1960), in his study of urban perception and design, emphasizes the importance of legible design characteristics which make a city easy to assimilate into consciousness. He calls this quality "imageability" and defines it as:

That quality in a physical object which gives it a high probability of evoking a strong image in any given observer. It is that shape, color, or arrangement which facilitates the making of vividly identified, powerfully structured, highly useful mental images of the environment.

A highly imageable (apparent, legible, or visible) city in this peculiar sense would seem well formed, distinct, remarkable; it would invite the eye and the ear to greater attention and participation. The sensuous grasp upon such surroundings would not merely be simplified, but also extended and deepened. ... The perceptive and familiar observer could absorb new sensuous impacts without disruption of his basic image, and each new impact would touch upon many previous elements. He would be well oriented, and he could move easily. He would be highly aware of his environment. (9–10)

Although Lynch relates imageability to physical structures, it is just as usefully applicable to designed, traversable spaces like the virtual environments in which digital games take place. We could take this one step further and claim that it is even more important in the case of environments whose representation is primarily visual and auditory. Lynch's definition of imageability can be modified by applying it not only to physical but also to virtual objects. In the context of this model, the most crucial part of Lynch's description of imageability is the greater ease of involvement that results from the absorption of spatial characteristics into consciousness in the process of updating the mental image of the space.

The two worlds featured in the qualitative research occupy opposite ends on a scale of imageability.[3] The continents in *Planetside* (Sony Online Entertainment, 2003) are more difficult to take in and distinguish, while those in *World of Warcraft* (Blizzard Entertainment, 2004) have been described as being easily learned and recalled. *World of Warcraft*'s regions have very distinct palettes as well as distinct types of vegetation, buildings, and terrain:

Blizzard did a good job in making them distinct. They vary them by color palette and by textures. Silverpine is blues and greens, with a lot of wood textures and fog. Barrens ... reds and oranges with rock textures. Stonetalon ... same color scheme as barrens, but with more forested textures. Arathi and Hillsbrad ... mostly greens with grass and leaves. Alterac ... white and snow, that kind of thing. Also, shapes. Silverpine ... lots of verticals, that kind of thing. (Rheric, *World of Warcraft*)

If we consider the coherence of Azeroth's terrain in *World of Warcraft* from a top-down view, it becomes evident that its designers were more concerned with creating distinct regions than having a realistic geographical layout, for tropical regions are placed adjacent to regions whose flora and fauna are indigenous to temperate climates. No explanation for the placement of regions is given, and none is required by players, as they are more concerned with the aesthetic qualities and imageability. *Planetside*'s Auraxis, on the other hand, makes more geographical sense than Azeroth, but players I interviewed felt less of a connection to its topography because its lack of imageability and aesthetic attraction made it hard for them to familiarize themselves with its landscapes. They expressed frustration with the difficulty they had remembering some of its regions, which naturally led to a sense of disconnection from the spatial dimensions of the game: "Most of the landscape isn't too different on a continent. All the trees look the same, etc. ... If I've been fighting in one place long enough I'll get the lay of the land. But a week later I'd be just as lost as before. Some areas though have stuck in my memory" (Kestra, *Planetside*).

Spatial involvement therefore describes not only the internalization of space immediately visible on the screen but also the ability to locate that space within a larger game area, often with the assistance of in-game maps

Figure 5.4
From map to three-dimensional environment in *The Elder Scrolls IV: Oblivion* (Bethesda Softworks, 2006).

or directions from other players (particularly in the case of MMOGs). Making the transition from the indexical map to the three-dimensional environment surrounding the player involves a distinction between inhabiting the environment and a more detached disposition which assists in learning the geography of the game world. Increased knowledge of the world makes it feel more familiar and easily navigable, while the freeing of attention resources allows players to focus on other aspects of the game, which often results in a deeper involvement.

The Space of Miniatures

The preceding discussion was primarily pertinent for game environments that embody the player in an avatar from the first- or third-person perspective. While avatar control games locate the player at a specific point in the environment, the omniscient perspective in turn-based and real-time strategy games as well as simulation games allows the player's point of view to wander freely above the landscape. Exploration may still take place if parts of the miniature environment are hidden from the player's view. Space is experienced in miniature, much as it is in dollhouses and tabletop war games. Miniature space disallows systemically upheld habitation: Players are always on the outside looking in, directing the action of multiple miniature entities. This disembodied relationship to space creates a radically different form of involvement, since players are not concerned about threats to their avatars located in the environment. In miniature space one can never be lost. Exploration in miniature space is an unraveling of a pre-existing picture without the danger of being exterminated or losing one's way. At worst, some of the controlled miniatures might be killed by enemies in the exploration, but more can be generated and sent out to resume exploration. In miniature environments, agency is tied not to a single in-game entity but to a conjunction of the capabilities of the miniatures under one's command and the player's disembodied powers. In *Empire: Total War* (Creative Assembly, 2009), for example, a player can order her troops to march across Portugal while she changes the tax rate in Silesia.

In such games, space can be drawn as a miniature landscape, as would be the case in most RTS games like *Warhammer 40,000: Dawn of War II* (Relic Entertainment, 2009). The miniature landscape is a representation of a particular environment built to support the interaction of miniatures within it. If the game involves warring factions of miniatures, then the environment incorporates defendable positions, cover for troops, and lines of fire dependent on the speed of movement, weapon ranges, and other characteristics of the miniatures that will inhabit it.

Figure 5.5
Miniature landscape in *Warhammer 40,000: Dawn of War II* (Relic Entertainment, 2009).

Many miniature landscapes in games contain resources that are desirable for the attainment of game goals. These might be locations where resources can be mined, special items that confer an advantage to the faction owning them, or areas where specific structures can be built. Miniature space tends to be designed with tactical goals and affordances in mind. The isometric view does not need to afford interesting on-the-ground exploration and navigation even if the landscape is rendered in three dimensions and could, theoretically, embody the player in an avatar wandering through it. Understanding how to find and control resources turns the miniature landscape into a tactical space. Internalizing space in miniature games thus becomes an exercise in understanding and exploiting the tactical potential of certain locations. While this is also true of avatar control environments, miniature landscapes tend to emphasize the capture and management of resources more heavily.

Some miniature environments are purely symbolic in nature. These tend to be puzzle games and digitized versions of board games or close derivatives. Here, spatial involvement is directly related to spatial control. *Risk* (Hasbro, 2009), for example, focuses on territorial control as its central mechanic: Players' capabilities are directly related to the territories they control. Similarly, in more abstract board games like checkers and chess, controlling space is key to winning the game. In these instances, space is not a habitable landscape, but a resource that requires control for competitive purposes. In these cases, spatial involvement becomes inextricably coupled with tactical involvement, both from the experiential and design perspectives.

The spatial-involvement frame defines players' engagement with the spatial qualities of a virtual environment in terms of spatial control, navigation, and exploration. It accounts for the process of internalizing game space, which is a powerful factor in engaging players and giving them the sense that they are inhabiting a place rather than merely perceiving a representation of space.

6 Shared Involvement

Macro Phase: Socializing in Game Worlds

Prior to their digital incarnation, games generally required multiple players and were often played in public spaces. The rise of digital games resulted in an increased potential for the creation of games in which players would be pitted against an automated system rather than human opponents. While nondigital games like solitaire, the *Fighting Fantasy* (Livingstone and Jackson, 1980) solo-RPG book series, and more complex strategy games like *Wolfpack* (Simulations Publications, 1975) already allowed for solo gaming, computers were the perfect vehicles for working out the calculations required to play them. They also eventually provided graphical representations of the games, serving to replace the more expensive, though perhaps more charming, miniatures and model terrain.

Earlier digital games, such as *Spacewar* (Russell, 1962) and *Pong* (Atari, 1972), were designed to be played by two players, and this trend continued through every generation of home computing technology. From the Atari VCS to the Nintendo Wii, gaming machines have been built with more than one player in mind. PC gaming, however, was not as amenable to multiplayer games until the advent of networked, multiplayer gaming, first over the Internet and later via local area network (or LAN) parties. LAN parties consist of a number of players connecting their machines through a router in the same building, often in a room rented specifically for the occasion. In most countries, LANs are organized on a regular basis, and often include competitive tournaments.

Today, on-line multiplayer gaming is becoming increasingly popular. Aside from games embedded in social applications like Facebook, the current generation of consoles have integrated multiplayer play and communication as part of their basic interface. With the PlayStation 3, Sony went even further by developing *Home* (SCE London Studio, 2008), a virtual world that players can access whenever they switch on their PlayStation 3. Any gamer with an on-line account is given a customizable avatar and an apartment they can decorate to their liking. Players can display trophies they have earned in games and also invite friends to their apartment to game or watch videos together. Although not a geographically continuous space like most MMOGs, *Home* is characteristic of the direction gaming is taking. PCs and consoles have become gateways to a vast network of other players with whom one can play, compete, and socialize.

One of the main attractors of MMOGs is their ability to support the presence of multitudes of globally distributed players in one environment (Steinkuehler and Williams, 2006; Taylor, 2006). This social participation can take many forms: direct conversations in group chat channels, the presence of other human-controlled characters one can bump into while traveling, or simply the background hum of general chat.[1] Even when players are not collaborating directly with others or competing against human opponents, the presence of their avatar creates a broader engagement with the world community. The presence of other players as a potential or actual audience considerably modifies how we conceive of and interact with the game environment. MMOGs offer more than just a game system or automated game world to interact with: They are built to foster persistent social structures that create a sense of being in a living, breathing world. As one research participant put it: "I think what really strikes me is knowing that all these thousands of characters running around are actually people stuck to their PCs all over the world. There through my monitor is a whole living world" (Baal, *World of Warcraft*).

What sets MMOGs apart from other digital games is this propensity to become full-fledged on-line societies. Friendships are started and ended, relationships formed, marriages sealed, and betrayals played out. An excellent example of the latter is the notorious heist within *EVE Online* (CCP Games, 2003) in April 2005. A mercenary corporation[2] known as the Guiding Hand Social Club was hired by a player-run corporation to infiltrate and sabotage an enemy corporation, Ubiqua Seraph. After months of careful planning and assimilation into Ubiqua Seraph, Guiding Hand made away with over US $16,500 worth of virtual assets and brought Ubiqua Seraph to its knees.[3]

Figure 6.1
A wedding inside *Second Life* (Linden Lab, 2003). Image courtesy of Marco Cadioli
(a.k.a Marco Manray).

The social ties and tensions formed in a virtual world are not separate
from the everyday; in fact, part of MMOGs' popularity arises from the ease
with which social conventions developed in other contexts cross over into
the game. As a number of research participants stated, once characters
attain the highest levels in the game and the bigger instances[4] are run more
than a few times, players often log on to retain contact with friends made
inside the world. Geography permitting, these friendships can turn into
out-of-game relationships: "Two of the three people I live with are people
I met through a MUD [multi-user domain]. I moved cross-country to move
in with them, actually. Came across on a three-week visit and decided to
just get a job and stay. The friendships are really fulfilling" (Splate, *World
of Warcraft*).

Having common interests is an important basis for many friendships,
and so it is not surprising to find that players sometimes feel they can
relate to others in the MMOG more than they can with people in their
daily life. At the very least they share one engaging interest: participation
in a virtual world. This shared interest is facilitated by the fact that MMOGs

are often designed to make collaborating with others necessary at certain stages of the game. Most importantly, joining a guild becomes a prerequisite for experiencing the so-called end-game content.[5] Being part of a guild provides the player with a regular group of people to interact with over a long period of time, increasing the potential for deep relationships to develop. Doing things with others creates a social bond that often exceeds the intensity of experience of other forms of on-line communication, such as chatting. Guild members must depend on and protect each other while overcoming other players and computer-controlled agents together.

On-line communication is thus contextualized within a history of shared experiences that give more meaning to a specific interaction than a discussion over a chat channel ever can. On the other hand, chatting about things outside the game occurs frequently, particularly when players are engaged with easy or repetitive tasks like grinding or harvesting materials. Traveling is also a great opportunity to converse with others, as long journeys often have to be undertaken and frequently involve long stretches of traveling when players have little more to do than make sure their character is heading toward the right destination.

Some guilds are organized by geographical location, such as a guild in *World of Warcraft* (Blizzard Entertainment, 2004) made up of people from Austin, Texas, that one of the participants belonged to. The guild organized monthly gatherings outside the game, giving people a chance to see the faces behind the avatars they are used to playing with. Actions and communication in the game are thus given additional meaning since players know they might meet the people they are interacting with:

I tease one guy online all the time about sleeping with him cause I know he's recently divorced. I think he likes the attention. He's coming to a guild party here in Texas in two weeks so I will [meet him] then. We use Ventrilo during quests so I even flirt with them over that too. Actually I flirt less over Ventrilo cause when you attach a voice to someone, they become more human to you and less like the person you've invented in your head. (Nombril, *World of Warcraft*)

Like other forms of on-line communication, MMOGs offer ways of being social that might not be possible in daily life for geographical, personal, or community reasons. If, as Edward Castronova (2005) argues, MMOGs are the current instantiations of the virtual reality worlds heralded in the 1990s, the reality they create is not what is generally meant by references to *the real*, that is, tangible, physical matter; rather, they create the social reality of everyday life with all the opportunities and problems this involves. MMOGs also differ from other forms of on-line communication in that the goal-directed activity they encourage creates a context and motivation for communicating and collaborating with others (Klastrup, 2004; Taylor, 2006):

You have the opportunities to get to know other people and make new friends. ... Chat rooms on the Internet don't have a "goal." MMORPGs offer the chance to work together at various puzzles, as opposed to just typing back and forth. (Haelvon, *World of Warcraft*)

Governance in Massively Multiplayer On-line Games

In late January 2006, a new guild in *World of Warcraft* announced its formation. As part of the process of describing the nature of the guild, the founding member, Sara Andrews, stated that the guild would welcome gay men, lesbians, and transsexuals. When some players heard the announcement, they reported Andrews to Blizzard Entertainment, who then warned Andrews to stop the guild-forming process or face having her account suspended. They claimed that Andrews's guild violated *World of Warcraft's* Terms of Use, under "Harassment—Sexual Orientation":

When engaging in Chat in *World of Warcraft*, or otherwise utilizing *World of Warcraft*, you may not ... transmit or post any content or language which, in the sole and absolute discretion of Blizzard Entertainment, is deemed to be offensive, including without limitation content or language that is unlawful, harmful, threatening, abusive, harassing, defamatory, vulgar, obscene, hateful, sexually explicit, or racially, ethnically or otherwise objectionable, nor may you use a misspelling or an alternative spelling to circumvent the content and language restrictions listed above. (*World of Warcraft: Terms of Use 2006*)

The offense, in this case, was that Andrews's guild brought to light the sexual orientation of its members, even if this was done only in the capacity of being "friendly" to gay men, lesbians, bisexuals, and transsexuals. The ironic aspect of the dispute is that Andrews felt the need to be explicit about the guild's policy on sexuality in reaction to Blizzard's inability to uphold their Terms of Use whenever the word *gay* was used, as it often is, in general chat as a derogatory term. Because the word was used so frequently in public chat channels, it was easier for Blizzard to corner one entity that brought the issue of homosexuality to light than to monitor its chat channels for such infringements. Blizzard adopts a self-policing policy, where members are asked to report instances of policy infringement. Thus, in a practical sense, what is acceptable is ultimately dictated not by Blizzard but by the general consensus of the population of the server. If there are too many players using a derogatory term, it becomes impossible to report them all. Blizzard addressed the controversy with the following statement:

To promote a positive game environment for everyone and to help prevent such harassment from taking place as best we can, we prohibit mention of topics related to sensitive real-world subjects in open chat within the game, and we do our best to take action whenever we see such topics being broadcast. This includes openly advertising a guild friendly to players based on a particular political, sexual, or religious preference, to list a few examples. (Blizzard, cited in Hunter, 2006)

In a personal communication to Andrews, Blizzard claimed, "Many people are insulted just at the word 'homosexual' or any other word referring to sexual orientation" (para 8).

This incident serves to illustrate just how dangerous notions such as the magic circle become when used to analyze or make decisions in virtual worlds like MMOGs. Blizzard has repeatedly stated that, within the confines of the game, communication should be limited to discussion relating to the game and should not gravitate to political or social issues, for fear of breaking the pleasure of the game for other players. For Blizzard, upholding the notion of the magic circle is an attractive option, as it enables the MMOG to be treated as a game cordoned off from the social concerns of the everyday and the problems of governance these entail. Companies like Blizzard expressly design gameplay elements to foster social ties, with the knowledge that such ties create a lasting connection to the community, which keeps users playing and thus paying monthly fees. Building community helps to create richer gaming experiences and, more importantly from a business perspective, it can often supplement gameplay with social participation that absorbs players, decreasing the need of scripted content. When the lure of game goals and achievements starts to wane, the social aspects of play can extend interest in the virtual world long enough for the developers to release new areas to explore and other game content to renew the players' interest.

Playing in a virtual world with others is a crucial motivator for players to try out and then continue to participate in such games. This section has focused mostly on MMOGs because they are the gaming environments most dependent on the development of societies of considerable size in order to exist at all. Not surprisingly, when asked which aspects of MMOGs interested them the most, the majority of participants replied that it was the sense of being part of a larger, evolving community that attracted them: "I'd have to say that the thing I like the most is the ability to engage in a vibrant online world with other players. The shared social experience of working together to do things is great" (Inauro, *World of Warcraft*).

Participants often discussed the distinction between their sense of community within an MMOG and in other multiplayer game lobbies and Internet chat rooms. The persistent geographies of MMOGs allow for a feeling of being surrounded by others because one can actually see representations of the community members wandering around the world. There is a particular kind of excitement associated with the potential to meet someone new with an ease that can be hard to come by in everyday life, particularly in the context of some cultures and geographical locations. In his book on social spaces in the contemporary United States, Ray Oldenburg (1999) notes that "the essential group experience is being replaced by the exaggerated self-consciousness of individuals. American lifestyles, for all the material acquisition and the seeking after comforts and pleasures, are plagued by boredom, loneliness, alienation" (13). Constance Steinkuehler and Dmitri Williams (2006) used two sets of ethnographic data from participants in *Lineage* (NC Soft, 1998), *Lineage 2* (NC Soft, 2003), *Asheron's Call* (Turbine Inc., 1999), and *Asheron's Call 2* (Turbine Inc., 2002) to examine whether MMOGs fit Oldenburg's criteria of "third places": "Most needed are those 'third places' which lend a public balance to the increased privatization of home life. Third places are nothing more than informal public gathering places. The phrase 'third places' derives from considering our homes to be the 'first' places in our lives, and our work places the 'second'" (Steinkuehler and Williams, 2006).

Steinkuehler and Williams conclude that MMOGs fit the majority of Oldenburg's (1999) criteria for third places. They also state, however, that MMOGs are best at creating broad social connections between people from different backgrounds ("bridging"), rather than promoting deep, emotionally supportive ones ("bonding"). Yet, since Oldenburg's argument is mostly aimed at the lack of opportunity for building the former type of social capital, Steinkuehler and Williams conclude that MMOGs are indeed sites which can perform the function of third spaces that Oldenburg feels are so sorely needed in the contemporary, suburban United States.

The attractions of playing with others is thus a salient factor in the current boom of virtual worlds, and if Steinkuehler and Williams are correct, this popularity is further fuelled by a decline in the quality and accessibility of off-line third spaces. The issues discussed by Steinkuehler and Williams are based on the United States, however; the social needs that MMOGs address may not be as powerful in other societies that have retained the prominence of third spaces within everyday social life.

Micro Phase: Collaboration, Competition, and Cohabitation during Gameplay

Literature and film are populated by characters of various levels of complexity. Some characters develop, others remain flat throughout. The writer can decide to place certain characters in leading roles through extended parts of the work, while other characters play a more minor, supporting role. And of course there are the rest of the agents that populate the world and maybe intervene in a scene or two, often in a functional manner: the newspaper salesman, the waitress at the late-night diner, and so on. The world of a literary work or film is always populated by characters that have been outlined and presented in a certain way by the writer. To this, the reader adds her own interpretation: She might even add information about the character or read in things that were not intended by the writer. In any case, we can say the literary and film characters have a life of their own based on the characteristics and histories that the author has decided upon for them.

Game environments are also populated with agents, but, unlike in literature and film, these agents are not merely described to us through representational means. Along with their representation they have coded properties such as behavioral patterns, psychological states, and physical abilities. The pack of one-eyed rhinos in *Spore* (Maxis Software, 2008) that attacks our peaceful group of grass-munching ostrich-like creatures does so not because the developer decreed that they would, but because those one-eyed rhinos have a coded behavior that dictates that they attack weaker creatures when they come within a certain distance. These properties are not just described or implied by the author but constitute the very machine that animates the on-screen representation into action. They are determined by the combination of the framework created by the designers, the creative touch of the player who made the creatures, and the environmental circumstances the creatures found themselves in. An entity in a game is thus not only a flat or rounded character by description and implication, but also a simulation with certain characteristics and behaviors.

As we have seen in the section on the macro phase, however, this is not the whole story. Agents in game environments can also be human-controlled. As a result, characters are like improv theater actors set in a larger environment than a stage. The avatars seen running around you in *World of Warcraft*'s Ironforge city are other players acting on their needs and whims. When interacting with these characters, you know you are interacting with other humans through your avatar: "It somehow makes the experience more immersive/intense knowing that there's a real person behind the avatar" (Sunniva, *Planetside*).

How people understand and react to other agents in the game world tends to depend on whether they think the agents are controlled by humans or by a computer. At times, there can be ambiguity between the two, especially when a player is new to a particular game or genre of games. But generally, players will understand which agents are being controlled by other humans and which are not. Once this distinction is made, the player's disposition toward the entity tends to change. Broadly speaking, there are four dimensions of shared involvement: cohabitation, competition, and cooperation.

Cohabitation

The presence of other agents in the environment can transform a virtual environment from a place that is dreary and empty to a place that is alive and interesting. The appeal of cohabitation is particularly relevant in the case of avatar-based games, as it anchors the player firmly to the location both spatially and socially. When a player walks down the street of a city in a game environment that is full of agents walking, talking, and interacting, she will tend to get the feeling of being in a living city, obfuscating the fact that she is interacting with a simulation. This impression tends to be even stronger if the agents are not static or obviously patterned. Human players offer the most potent sense of cohabitation, as the player is aware that there are other human minds sharing the experience of that street (Lim and Reeves, 2006).

Humans instinctively perceive computers and computer-controlled agents as social beings. Byron Reeves and Clifford Nass (1996) have found that people treat media interfaces as social actors at the slightest provocation, from a request by the computer for help to compliments from the computer. An interesting finding in Reeves and Nass's research is that our first reaction to media representations is to treat them as if they actually existed:

Social and natural responses do not require strategic thinking; rather, they are unconscious. The responses are not at the discretion of those who give them; rather, they are applied always and quickly. ... The automatic responses defined by the media equation can be initiated with minimal cues. We saw the equation work with personality expressed in a couple of words or with a simple line drawing. (252–253)

These findings were corroborated by Gerhard, Moore, and Hobbs (2004), whose research indicates that players derive a sense of cohabitation in a virtual environment even when the computer-controlled agent in the virtual environment is rudimentary. If such minimal cues were sufficient to give participants in Reeves and Nass's experiments a sense of interaction with a social entity, we can assume the same to be true of computer-controlled game characters, particularly since these are communicated with a higher quality of visual and aural representation and demonstrate more complex behavior. As M. Ryan Calo (2010) argues, this tendency is only growing with technological developments in computer-controlled agents' behavior and their visual and aural representation.

When Dimitri Rascalov betrays the player's character, Nico Bellic, in *Grand Theft Auto IV* (Rockstar North, 2008) after Nico has risked his life killing Dimitri's previous boss on his request, players experience the event as betrayal by a double-crossing mobster rather than as an interaction with a computer program (Gerhard, Moore, and Hobbs, 2004; Reeves and Nass, 1996; Ruvinsky and Huhns, 2009). Thus, although there is a distinct difference between players' relationships to human avatars and computer-controlled agents, both have the potential to generate a sense of cohabitation in a virtual environment that is described by this dimension of the model. An added dimension of shared involvement is provided by persistent MMOGs.

Figure 6.2
Surrounded by AI characters in *Assassin's Creed* (Ubisoft Montreal, 2007).

In MMOGs, characters exist over an extended period of time, with no option of changing a character's name or appearance aside from altering his or her clothing and equipment. Consequently, players accumulate a reputation, positive or negative, among other players on the same server. Some guilds or outfits keep "kill on sight" lists for players whose actions are deemed to be unacceptable, such as *ninja-looting* (using unfair methods to take rare items off killed mob members and then leaving the group) or *ganking* (killing characters that have no chance of winning a fight because of a difference in levels, for no particular gain). In MMOGs, therefore, one's social standing tends to have a significant impact on the decisions players make during actual gameplay.

A further element related to cohabitation is spectatorship. In the case of multiplayer games, particularly those that allow the ability to look through other players' perspectives when one is not playing, one's actions become a performance watched and often commented on by others. Players of on-line multiplayer games tend to keep the same alias across different games in order to be recognized within a particular gaming community, thus building reputations based on their actions. Joining a clan[6] usually requires submitting to a trial in which applicants are screened by senior members who assess the applicants' abilities by following their actions in spectator mode for a number of game rounds.

Spectators can also be located in the same physical space as the player. The scenario of playing on a living room console is a good example. If you are playing *FIFA 2009* (EA Sports, 2008) against a friend while being watched by a group of others, your awareness of the audience, cheering for you or your opponent and gasping at shots that hit the posts in the last few minutes of a drawn game, has the potential to involve you more deeply in the game and intensify your affective involvement. In this way, your actions, both in the game and in the physical environment, become a performance for an audience. This performer/audience relationship can also prompt players to perform in a way that is more appealing to the audience, thus shifting the focus from functionally goal-directed behavior to more visually pleasing or impressive actions.

Cooperation

Another aspect of multiplayer gaming is the importance of collaboration to achieve common goals (Manninen and Kujanpaa, 2005; Heide-Smith, 2007). The need for collaboration creates the potential for involving players through communication and teamwork. *Halo 3* (Bungie, 2007), for example, can be played cooperatively by two players on a split screen in the same living room. The pleasure of these so-called co-op games in the same physical location lies in the immediate interaction players have with each other, making it easier to coordinate their efforts both verbally and by seeing each other's screen. If one player is lost, for example, the other player can more readily direct him to a meeting spot or go and find him in-game than would be possible if the lost player had to describe where he was over voice chat. The main limitation for co-op games played on a single machine tends to be on freedom of spatial navigation: Players are often limited to staying in close proximity to each other and levels are designed to suit this need, making them linear and spatially restricted.

Another form of cooperative play in the same physical location is LAN gaming, which takes place on a number of computers connected to each other in the same room or hall. These events are sometimes held privately, in players' homes, or at larger venues, often with hundreds of players participating. Larger LANs often host tournaments and act as the main hub for competitive play. At this level of competition, teamwork requires a state of seamless communication and cooperation that can be seen as the pinnacle of real-time, networked collaboration.

Games designed primarily for cooperative play have been growing in popularity in networked PC gaming. The most striking example of this type of game is *Left 4 Dead* (Valve, 2008), which places players in the scenario of a group of postapocalyptic survivors who have to make their way to a rescue point through a zone infested with zombies. There are four survivors, each ideally played by a separate player. The players need to help each other and stick together in order to get through the perilous environment. The game uses an AI director to manage dramatic tension by spawning mobs of zombies, not only in specified locations, but also when players separate from their group or take too long to traverse an area. Besides garden-variety zombies, there are four special zombies, each with an ability which, when successfully unleashed on a player, requires the help of his teammates to overcome. One of the zombies, called the Hunter, attempts to pounce on players. If he successfully lands on a player, she is immobilized and starts losing health rapidly as the Hunter claws and bites at her. She cannot do anything to liberate herself from her leaping assailant and needs a teammate to shoot the Hunter off. Similarly, the Smoker extends a long tongue that wraps itself around the player's neck and draws her toward him. She has a few seconds to try to shoot him off her, but after that is dragged and slowly constricted. Once again, her teammates must come to save her if she is to survive. Aside from these mechanics which foster interdependence in order to survive, the game also facilitates collaboration by allowing players to see each other through walls, giving a sense of where one's teammates are at all times. Here, realism is traded for a visual device that ensures players do not get lost in an area, allowing them to stay together more easily. Players also have a medical pack that they can use only once, either on themselves or on others. With the sheer number of opponents and limited ammunition supplies, the game can be quite challenging, further adding to the need for collaboration and clear communication.

In the realm of MMOGs, teamwork is essential at various levels. Players can help each other out in an ad hoc fashion throughout the world. If you see another player being chased by wolves, you can step in and try to fend them off or at least distract enough of them so that you and the other player can take them on together. This potential for spontaneous collaboration can foster social relations based on more than idle chat. A number of research participants commented on the satisfaction of being able to help others and cooperate with them in this ad hoc manner.

MMOGs build their subscription based principally on collaborative play and are specifically designed for players to team up together on two levels: groups and guilds. Grouping is a necessary aspect of most MMOGs, because the games are designed in such a way as to require teamwork as characters progress in levels. In most MMOGs, players can undertake quests or missions on their own or in groups. As the character progresses in skill, which is usually measured in levels, quests require players to form groups in order to complete the levels, effectively requiring teamwork. At later stages, the small five- or six-person groups become groups of fifteen, twenty, or at times even forty people that invest a good number of hours in completing a set of quests, usually within a self-contained scenario. Due to the investment of time and the need for functional collaboration and a considerable amount of preparation, players need regular teammates who will work together not only during the quest itself but also in preparing for it. This encourages, if not compels, players to band together in guilds. Guilds facilitate cooperation by having players communicate over a common chat channel, enabling them to get to know each other over time. Third-party voice chat programs such as TeamSpeak or Ventrilo are often used to supplement the chat channel. Guilds also organize virtual social events and minigames intended to bring guild members closer together and to ensure a fun atmosphere for all.

Participants stated that the possibility of working with other, geographically distant people to reach a common goal is an important, involving factor in MMOGs: "Hundreds of people 'attempting' to work together for a common goal in a FPS. The idea of virtual armies that have a never-ending battle really appeals to me. It's also the coordination of your specific allies to meet a certain goal. You get all warm and fuzzy when your team is kicking ass" (Bladerunner, *Planetside*).

The cooperative aspect of shared involvement seems to become more engaging the greater the number of people working together. Much can go wrong, but, when the collaboration works, the efforts are seen as being more than worthwhile. When participants were asked to relate memorable

Figure 6.3
Players from the Ghosts of the Revolution outfit in *Planetside* (Sony Online Enter-
tainment, 2003) form up for a massed attack. Image courtesy of Ghosts of the
Revolution.

sessions, a large percentage, particularly among players of *Planetside* (Sony
Online Entertainment, 2003), described situations of successful mass collaboration in large battles. Curiously enough, even though participants in
World of Warcraft (Blizzard Entertainment, 2004) included players from
player-versus-environment (PvE) and player-versus-player (PvP) servers,[7]
the instances of involvement through collaboration mentioned were all
related to PvE situations, particularly in the case of the higher levels. All
but two participants stated that the main reason for their participation in
MMOGs was interaction with others, whether collaboratively or competi-
tively (although the collaborative aspect was cited most frequently as a
motivation).

Competition and Conflict

The competitive element of games, referred to by Roger Caillois (1962) as
"agon," has been around since their inception. Digital games have not
decreased this penchant for competition; if anything, they have amplified

the potential to encompass thousands of simultaneous players on each side of a contest that could span participants from all over the world. This ability to compete against a multitude of globally distributed players in real time constitutes a significant paradigm shift in the scope and potential of competitive activities.

Digital games offer a variety of forms of competition, ranging from traditional tabletop games like cards, board games, and strategic war games, to simulations of physical contests like football, basketball, and other sports, to simulations of nonsportive contests like gunfights, aerial dogfights, or medieval sword fighting. A key attraction is that players can compete in any number of historical or fantastic contests that either are not possible in everyday life or have dangerous or terminal consequences.

Early digital games pitted two players against each other. *Pong* (Atari, 1972) is an abstraction of tennis, while *Combat* (Atari, 1977) was a collection of two-player games simulating tank and air battles in a similarly abstract fashion. Although simple, these games proved to be fun because they set the stage for synchronous competition between players.

Today, the possibilities for competition are vast. Players can go head to head on the same console in split-screen car-racing games like *Gran Turismo* (Polyphony Digital, 1998) or on the same screen, as in *FIFA 2009* (EA Sports, 2008). There are also situations on consoles where players need to work together to complete a task but are simultaneously competing against each other for the better score, as in *Rock Band* (Harmonix, 2007).

In the case of direct competition, the competing factions aim for the same goal, but conflict can also occur among entities in the game world with differing objectives. A plan to travel between the cities of Bree and Rivendell in *The Lord of the Rings Online* (Turbine, 2007) will lead to a conflict with computer-controlled creatures that assail you on approach. Similarly, in attempting to complete quests in the Angmar region of the same game, you will likely face conflict from groups of player-controlled creatures focused on another objective: capturing areas from opposing bands of players. Sprawling virtual environments can therefore create situations in which conflict does not even have the same goal, but is either a facilitator or an obstacle to reaching one party's goals.

Computer-controlled agents provide the primary source of conflict in single-player action games like *Half-Life 2* (Valve Software, 2004),

S.T.A.L.K.E.R. (GSC Gameworld, 2007), and many others. Although not as adaptable or unpredictable as human players, computer-controlled agents are becoming more sophisticated and are already able to learn from players' actions and act accordingly. This development has broken the highly predictable patterns that made them less engaging opponents in their more primitive forms such as in *Space Invaders* (Taito, 1978).

Malaby's (2007) emphasis on contingency is helpful here. One aspect of Malaby's model of games as domains for the interaction with contrived contingency is "social contingency":

> The unpredictability of never being certain about another's point of view (and often, resources) is a key component of games such as chess, poker, and countless others. The extent to which (economic) game theory has focused on differences in information is a reflection of the correct recognition of social contingency as a factor in games, but it is never the only source of contingency. It is not simply the challenge of making the right guesses about others' points of view which is involved, it is acting on those guesses. (Malaby, 2007, 16)

Malaby rightly points out that social contingency is codependent on performative contingency. Players tend to be engaged by the fact that they do not know the full extent of their opponent's capabilities and plans. At best they can make educated guesses about the likely course of action an opponent or group of opponents will take, and then act accordingly. This action is further calibrated according to the player's perception of her own ability in handling the challenge set by her opponents. Social contingency helps us understand why more competitive players prefer playing against human players instead of computer-controlled agents: "There are actually other people playing with you ... or against you. Not just some AI telling something what to do all the time. That, and there is just so much depth to them. It really just grabs you and takes you for a ride. The human element adds a lot of fun and surprises to a game. Makes it more challenging and realistic" (BunkerBoy86, *Planetside*).

Due to the current limitations of technology and the fact that computers act according to predefined rules set by their designers (although these rules can, at times, combine in ways that are not even predictable by those who created them), the social contingency of an AI agent tends to be mapped more easily by players who are attentive to its behavior. The attention to an opposing player's or AI agent's behavior in an effort to reduce one's social contingency is only possible once a player's attention is not

consciously engaged by other aspects of involvement, particularly kinesthetic involvement. If we are still establishing how to move around and
interact with the environment, we will not be able to make informed decisions about the plans and expected behavior of our opponents. Like the
other dimensions, shared involvement can require conscious attention or
can be internalized by a player. Both cooperative and competitive aspects
of games tend to require a good deal of conscious, or at least semiconscious,
learning to take place in order to reach proficiency and, eventually, mastery
and internalization.

The ability of computer-controlled agents to present a sufficiently challenging competitor depends on the game in question. In a strategy game
like *Empire: Total War* (Creative Assembly, 2009), for example, computer-
controlled factions present opponents who are adequately challenging
during the campaign mode but are more easily overcome during real-time
battles. The challenge for the player is that the AI-controlled units of troops
act simultaneously according to the behaviors that have been coded into
them, while the player has to manage each unit individually, or cluster
them in groups. On the other hand, the AI does not act in a cohesively
strategic manner when the whole army is considered, and shows a lack of
awareness of each individual unit's role in relation to the rest of the army
and the opponent. It becomes obvious to any observant player that the
computer-controlled units follow scripted patterns of action triggered by
specific prompts, and thus lack a more holistic awareness of the overall
conflict. Thus, a unit of skirmishers will start running away from a unit of
light cavalry only when it comes within a certain range, and not before.
A human player would likely have noticed the flanking unit of cavalry
approaching their skirmishers and moved them earlier to avoid being
overrun. The computer does not seem to have an overall vision of the
battlefield and its units.

In FPS games, computer-controlled agents can be dangerous because of
their pinpoint accuracy. It is rare that they will miss a shot or be blocked
by a low-lying wall. On the other hand, they are often easily tricked. Once
again, computer-controlled agents do not seem able to take into consideration the entire environment and are not able to react to devious traps set
by players. Lim and Reeves (2006) have investigated the difference in
player arousal and engagement between facing opponents that are perceived to be controlled by other human players and facing those controlled

by AI. The results showed that having human opponents increases both automatic and rationalized responses to gameplay.

In on-line multiplayer FPS games, like *Counter-Strike* (Valve Software, 1998), *Call of Duty IV* (Infinity Ward, 2007), and *Red Orchestra* (Tripwire Interactive, 2006), play centers on a contest between two teams. Although the contests take place in the same spaces (maps), players will engage for hours on end due to the collaborative and competitive potential. The competition in these games scales up from the casual encounter with strangers on line to dedicated tournaments run by gaming groups and the increasingly popular professional gaming circuits. At this level of competition, players may spend many hours each day practicing with their teammates and competing against other teams in preparation for tournaments.

In FPS games, the move from playing as an individual in a random group of people on line to playing in an organized, regular team tends to be marked by a shift in a player's perception of her skill in relation to others. As players improve, they may be approached by, or may choose to approach, an established clan. Clans are teams of players who train and compete together on line and at LANs. Some clans have very strict rules and require a process of trials to screen for skilled players, while others are more casual and have fewer barriers to entry. When a player joins a clan for the first time, she must change her playing style to function productively with other players. As Rambusch, Pargman, and Jakobsson (2007) observe:

The emphasis thus changes from *individual* to *team play* and skills such as good communication and the ability to adapt to changes in the clan's line-up and the opposing clan's strategies and moves become increasingly important. Players come to view themselves not only as individual players but also as team players who know that everything they do can also have impact on the other members of the team. Competing in local tournaments starts to shift the activity from leisure to (semi) professional work and once players have won their first tournament they want more. As a player pointed out, "When we won we thought that we could achieve more and we started to play more."

The satisfaction of successful teamwork becomes a strong attractor for players to continue playing at a competitive level. The demands that competition places upon collaboration bind the two together to form a deeply involving experience that influences both long-term engagement with the game described by the macro phase of shared involvement, as well as the moment-by-moment involvement during gameplay described by the micro phase.

Shared involvement concerns the engagement derived from awareness of and interaction with other agents in a game environment, whether they are human- or computer-controlled. The interactions have been discussed here in terms of cohabitation, cooperation, and competition. Shared involvement thus encompasses all aspects related to being with other entities in a common environment, ranging from making collaborative battle strategies to discussing guild politics or simply being aware of the fact that actions are occurring in a shared context.

7 Narrative Involvement

Macro Phase: Narrative and Experience

In game studies, the discussion of games and narratives has been dominated by two related questions: "Do games tell stories?" (Juul, 2001) and "Are games narratives?" (Aarseth, 2004; Eskelinen, 2004). In addressing narrative involvement in the context of the player involvement model, however, our approach is to consider how stories are experienced in games. Arguing that narratives are not particularly important to the gaming experience is a nontenable, normative assumption that predetermines how players experience game environments. On the other hand, attributing every aspect of the gaming experience to narrative is equally unproductive. The approach taken here acknowledges the formal aspects of game narratives but focuses primarily on the manner in which these formal features, along with other aspects of the game environment, shape players' experience of narrative.

A particular strand of narratology represented by theorists like Seymour Chatman (1978), Gérard Genette (1980), and Gerald Prince (1982) has characterized narrative as a form of retelling. This approach has been described by Marie-Laure Ryan (2006) as the "speech-act approach to narrative" (5). In this conception, the story is always told in retrospect, so that what the reader or audience is experiencing occurred in a past that is now fully determined. Ryan also points out that this strand of narratology requires the existence of a narrator for a narrative to exist. This results in a radical position on narrative that excludes visual media, drama, and other performance-based media from being acknowledged as narratives.

In his *Narrative Discourse*, Genette (1980) clearly states that his work discusses only literary fictional narrative. Prince makes a similar argument and openly excludes drama from his consideration of narrative. Genette formulates narrative in Saussurian terms:

> I propose, without insisting on the obvious reasons for my choice of terms, to use the word *story* for the signified or narrative content (even if this content turns out, in a given case, to be low in dramatic intensity or fullness of incident), to use the word *narrative* for the signifier, statement, discourse or narrative text itself, and to use the word *narrating* for the producing narrative action and, by extension, the whole of the real or fictional situation in which that action takes place. (27)

Literary narrative has its own complex structures that more than justify a focused study such as Genette's. It is particularly useful, within such a study, to distinguish between the story and the structure of its retelling. Narratology commonly distinguishes between the actual events that occurred in chronological order and the way these events are presented to the reader (Metz, 1974; Genette, 1980; Chatman, 1978; Prince, 1982; Bal, 1997). Although there is considerable disagreement within narratology about the characterization of narrative, the one thing that narratologists agree on is this distinction between the abstracted sequence of events, referred to as *story*, and the presentation of those events to the reader, most often called *discourse* (Culler, 1981).

This linguistics-based position has not been without criticism in the field of literary theory. One concern is that there is no actual story outside of the discourse that constructs it (Culler, 1981; Walsh, 2007). Nevertheless, one can see the analytic sense in distinguishing the order in which events are presented and the means by which these events are presented from the original, chronologically ordered sequence of events. Such a distinction presupposes the presence of an author who configures a story that is actualized during the reading process. The author therefore creates a preset structure for the consumption of his work.

This distinction is not particularly productive in the case of game environments. As Jesper Juul (2001) has pointed out, the difference between story time and discourse time is often negligible in games:

> It is clear that the events represented cannot be *past* or *prior*, since we as players can influence them. By pressing the CTRL key, we fire the current weapon, which influences the game world. In this way, the game constructs the story time as *synchronous* with narrative time and reading/viewing time: the story time is *now*. Now, not just in the sense that the viewer witnesses events now, but in the sense that events are *happening* now, and that what comes next is not yet determined. (para 34)

According to Juul, the lack of importance in the difference between story time and discourse time is a telling sign that games are incompatible with narratives. This further leads him to claim that *"you cannot have interactivity and narration at the same time.* And this means in practice that games almost never perform basic narrative operations like flashback and flash-forward. Games are almost always chronological" (para 35, italics in original).

The problem with this argument is that Juul switches between two dimensions of narrative in game environments: the story generated by the moment-to-moment actions within the game environment and the story that has been pre-scripted. In *Max Payne* (Rockstar Toronto, 2001), for example, flashbacks and flash-forwards are important parts of the scripted game experience. In *Call of Duty IV* (Infinity Ward, 2007), there is a single chapter which takes place a few years prior to the rest of the events in the game. On the other hand, the generation of events through the player's interaction is not predetermined. If this were the case, then games would lose their ergodicity. Although these two aspects of narrative in games are related, it is critical to distinguish which one we are referring to when discussing the subject. To this end, I will here make a distinction between the *alterbiography*, referring to the story generated by the individual player as she takes action in the game, and the *scripted narrative*, referring to the pre-scripted story events written into the game. On the moment-by-moment level of engagement, a player's interpretation of events occurring within the game environment and his interaction with the game's rules, human and AI entities, and objects result in a performance which gives game environments their narrative affordances. Interaction *generates*, rather than excludes, story.

Ergodic media do contain important story elements, contrary to Markku Eskelinen's (2004) argument, but the form these narrative elements take is not adequately addressed by classical narratology. It seems strange for Eskelinen to claim, "It should be self-evident that we can't apply print narratology, hypertext theory, film or theatre and drama studies directly to computer games" (36) and yet, in the very same paper, to build an argument against narrative in games based on claims developed by the most ardent of literary narratologists. His invocation of the assertions (which I've critiqued) by Genette (1980) and Prince (1982) that narratives require a narrator undermines Eskelinen's own claim that we need to rethink existing theories originating from other media in the context of games.

As Ryan (2006) has stated, the arguments brought forward by Eskelinen and Juul, namely that games cannot suggest stories, merely imply that games do so in a way that is different from literature and movies. This is not a negative claim for games. Quite the contrary: game environments have reached a sufficient level of sophistication that not only allows, but demands, a redefinition of classical notions of narrative.

Experiential Narrative

Theorists who have written positively about narrative and games have invariably discussed the experiential aspect of game narratives. While there are considerable differences in the way the issue is formulated, there are also some important overlaps, which we will explore here.

Celia Pearce (2004) makes an important claim about the comparison of games to literature and film when she states that "unlike literature and film, which center on story, in games, everything revolves around play and the player experience" (144). It might be more accurate to say that literature and film are "narrative through and through": there is nothing outside the narrative that is being retold. Game environments, on the other hand, are forms of designed experience which, although they may include story elements, are subservient to the overall experience of the player.

Pearce proposes a set of six narrative elements, or "operators," that may be found in games, which she calls experiential, performative, augmentary, descriptive, metastory, and story system. The first of these is a component of all games, while the other five occur in different combinations. Although Pearce's framework is notable for its acknowledgment of the importance of player activity in forming an ongoing story, it suffers from an overgenerality that makes it difficult to apply practically. Pearce uses the framework to describe the narrative aspects of games such as basketball, tic-tac-toe, and Battleship. It seems largely uninteresting to discuss the narrative of a game of tic-tac-toe. A narrative framework that claims to be constructively applicable to such a wide spectrum of activities and media objects as basketball, tic-tac-toe, Battleship, chess, and *The Sims* (Maxis Software, 2000) will be severely challenged to be analytically productive when applied to game environments with strong narrative elements.

Like Pearce, Katie Salen and Eric Zimmerman (2003) emphasize the experiential dimensions of narrative elements in games. Their book *Rules of Play* takes game design as its primary focus and, like other practicing game designers, Salen and Zimmerman take the presence of stories in games as a given. Reading through articles on Gamasutra, talks at the annual Games Developers Conference, and various game design books, it is evident that the central question for game designers is not whether games are stories, but how best to convey stories through games. In an *Edge* article ("The Making of Grand Theft Auto IV," 2008), *GTA IV* lead designer Sam Houser discusses how the dynamic system of that game's environment creates moments that feel like pre-scripted narrative events. As increased storage and processing power enable designers to create more complex game worlds, there is a steadily increasing emphasis on the potential to relate narratives that adapt dynamically to players' input. As with Salen and Zimmerman, the majority of these talks and articles by game designers focus on players' experience of narrative.

Salen and Zimmerman (2003) make a distinction between embedded and emergent narrative. The distinction is useful in understanding game narratives, particularly because the emergent narrative accounts for the systemic structures of games: "It is the dynamic structures of games, their emergent complexity, their participatory mechanisms, their experiential rhythms and patterns, which are the key to understanding how games construct narrative experiences. To understand game narratives, it is essential to analyze game structures and see how they ramify into different forms of narrative play" (381–382).

This description echoes Aarseth's (1997) discussion in *Cybertext*, which stresses the importance of taking into consideration the mechanical, coded structures of ergodic texts, not merely their surface signs. In order to develop a coherent and sustainable framework of narrative analysis to be used in the context of game environments, the emergent narrative to which Salen and Zimmerman refer needs to be anchored in the game elements that generate such a narrative. The major challenge here is to not let the experiential nature of this component of narrative become so general as to be analytically redundant.

Although it is essential to look at *how* games create stories, Salen and Zimmerman, like Pearce, stretch the notion of experiential narrative beyond its limit as a useful concept. They do this, first, by not making a distinction between abstract games, sports, and virtual game environments, and second, by including general thoughts relating to a game as part of the experiential narrative:

The dramatic tension of Poker, too, gains its bite from the uncertainty of outcome. Bluffing contributes to the narrativity of the experience, heightening the potential for deceit. As players enter into the psychological space of the bluff, narrative tensions mount. *Does she really have the hand she says she has, or is she bluffing? What if she isn't bluffing? Can she still be beaten? He just made a large bet, so he must have a good hand. But he bluffed last round, and he wouldn't try that same trick twice in a row.* (Salen and Zimmerman, 2003, 388; italics in original)

The importance of experienced narrative to a sound theoretical conception of narrative in game environments becomes problematic when we can apply the concept to any interaction with the game system or to any thoughts relating to it, as in the example just given. As Ryan (2006) has shown, a cognitive perspective on narrative can be applicable to a variety of media while catering for the specificities of the form of media object in question, but we must ground the experience in the (cyber)textual qualities that the narrative experience derives from. Salen and Zimmerman's poker example considers thoughts about other players vis-à-vis the state of the game as a form of narrativity. It is important to distinguish thoughts about the strictly ludic dimensions of a game system from experiences that have a story element related to the fictionality of the game environment itself.

There is a qualitative difference (at least in terms of narrative) between the poker example and the story generated through interaction with Liberty City's inhabitants and environment in *Grand Theft Auto IV* (Rockstar North, 2008). For example, after doing a favor for the protagonist's cousin Roman, we take our character Nico across the street for a hot dog. While eating, he receives a call from Michelle, a girl he had a rather unsuccessful date with a few evenings ago. It seems, however, that Nico is still in her good graces, for she asks him to meet her later that night. As Nico's life has been rather stressful lately, we decide to let him relax for the moment and go out with Michelle. Following a comment by Michelle about Nico's poor dress sense, we take him to a clothes store to pick out a new outfit. We then direct him to a deserted alleyway to find an appropriate vehicle to take on the date. As we turn a corner, we bump into an irritable man who insults Nico. Having a short temper (or an itchy trigger finger), we instinctively throw a punch at the offending party, failing to notice a police car patrolling the area. A siren wails and we send Nico running away down the narrow alley ...

As we play *Grand Theft Auto IV*, this narrative forms in our minds not only via the representational elements and underlying code of the game, but also through our subjective interpretations of them. Nothing in the game system tells us that Nico is tired, but after having failed the previous mission a few times and become frustrated with it, along with the fact that it was night and it had been a confusing week in a new country, we decided that Nico was hungry and exhausted. Similarly, our decision to punch the bystander was a blend of our own temperament and our interpretation of Nico's character, along with the desire to see what happens that is common to gameplay. Our ongoing narrative is not merely formed from a type of free association of generated events, however, as it is grounded in the game environment's reacting to our specific actions and the actions that the game system affords and encourages. Our interpretation of Nico's reaction to the insult and his subsequent violence result from a combination of imagination and interaction with the game's constituent elements. The action is manifested in the game environment at the levels of code and resultant representation. Salen and Zimmerman's example, on the other hand, is constituted of thoughts that, although related to the game system, are not manifest in a narrative manner within the game environment because there is no game environment that supports the ongoing generation of narrative. In the case of poker, the game does not include an environment inhabited by characters. The players interact with a stack of cards bearing a mixture of symbolic and iconic signs and a rule system that generates the game. The game system and its representational level do not support the generation of what I am here calling *alterbiography*.

Henry Jenkins (2004) also emphasizes the experiential nature of game environments in his paper "Game Design as Narrative Architecture." He argues that when considering the ways games create narratives, we should rely not on their temporal structures but on their spatial ones. He outlines four types of spatial narrative forms: "Environmental storytelling creates the preconditions for an immersive narrative experience in at least one of four ways: spatial stories can evoke pre-existing narrative associations; they can provide a staging ground where narrative events are enacted; they may embed narrative information within their mise-en-scene; or they provide resources for emergent narratives" (Jenkins, 2004, 123).

Jenkins's narrative types are not mutually exclusive. Evocative narratives, for example, are a form of intertextuality, drawing upon other media texts to convey narrative meaning in a metonymic process. In *Lord of the Rings Online* (Turbine, 2007), players need not be told what elves and hobbits are or who Frodo is. Tolkien's works and their rendition in cinematic form have already established a well-known history of that world and populated it with characters. Evoked narrative conveys intertextuality primarily through embedded narrative: story elements that are built into the game space. Embedded narrative is not conveyed directly to players but depends upon their engaging with it. Jenkins's formulation does not, however, differentiate elements of embedded narrative assembled in the mind of the player from those that are conveyed as discrete narrative chunks, such as back-story or ongoing events portrayed through cut scenes.

Jenkins's approach to the experiential dimension of game narratives is tied to specific qualities of game environments. Unlike Pearce or Salen and Zimmerman, he connects the experiential dimension with the spatial, ludic, and performative structures of the game. Jenkins is focused on the spatial qualities of narratives in games. Although this is by no means a fault of his paper, a thorough account of narrative in games needs to address further dimensions than the spatial. Below, I discuss two concepts that ground discussions of experiential narrative in the specific qualities of a game, foregrounding the subjectivity of the player as the core of these narrative qualities.

Micro Phase: Alterbiography and Scripted Narrative

Although the main focus of this chapter is the experiential aspect of narrative, some consideration must be given to the narrative that is packaged into the game by designers, the *scripted narrative*. The experiential aspect of narrative, the alterbiography, is often informed by the properties of the scripted narrative. We will therefore look briefly at scripted narrative, then describe the concept of alterbiography and end with a consideration of how these two combine in the context of the player involvement model.

Scripted Narrative

In the majority of game environments, there is a story the designers want to impart to players. This can range from a very simple introduction to the game world along with an indication of the higher-order goals designed into the game system to more complex narrative situations involving multiple characters and plot twists. In many cases, players can decide to engage with the entirety of the scripted narrative or focus on the more goal-oriented tasks that push the game forward.

In MMOGs the quests that players embark on usually communicate two things: the ludic goal of the quest and the scripted narrative that surrounds the quest. While some players skim the quest text to work out what they need to do to accomplish the quest, others engage specifically with the story aspects. Whether or not players are interested in the scripted narrative, they will inevitably engage with it at least on the most rudimentary level. This is the case because story-driven games couple progression (both in terms of game goals/rewards and spatial exploration) with advancement of the scripted narrative. Designers are becoming more careful to ensure that players do not need to engage with the full level of narrative to experience and complete a game, as was often the case in earlier adventure games. If one missed a piece of information in *The Secret of Monkey Island* (Lucasfilm Games, 1990), for example, it might become difficult or impossible to progress.

Aside from the structured progression through the intended story, scripted narrative is also delivered to the player through different channels, including cut scenes,[1] quick-time events,[2] objects in the world, dialog, and straightforward streams of verbal text.

Scripted Narrative Structures of Progression

Higher-order ludic goals assigned by the game system are often tied to key points in the scripted narrative. At times, narrative goals drive game goals, while at other times the opposite is the case: Game goals are invested with narrative elements. Whichever pulls the other, it is safe to say that in the majority of story-driven game environments the completion of a game goal either progresses the scripted narrative, links the player to another sequence of quests, or ends there.

In some game environments there is only one overarching scripted narrative to follow, even if the player's alterbiography varies from one game session to another. In other games this main story line branches at various predefined points. Still others offer the choice to engage with the main story line, roam in the game world freely, or engage in similarly sectioned secondary story lines, often in the form of side quests. These side quests can be linked to the main story line or exist independently. At other times the game world contains no main story line, but may have a number of serially organized side quests that exist independently or in relation to each other. Following are three brief examples of different structures of progression.

Halo 3 (Bungie, 2007) is a game with a linear progression of scripted narrative directly equivalent to the game objectives. When players reach a specific checkpoint, they are allowed to progress further in the game, and at certain junctures within that progression the scripted narrative of the game is advanced, usually through a cut scene or through communication with one of the game characters. The spaces in each level allow players to navigate and reach objectives in various ways, but the overall narrative progression is strictly linear.

The scripted narrative structure of *Fallout 3* (Bethesda Game Studios, 2008) contains a branching backbone quest and a number of side quests. *Fallout 3* takes place in an open environment within which players can roam. As a result, the selection of side quests that players come across depends considerably on their exploration. Because they have an open world to explore, the players' alterbiography will be more variable than those of linear progression narratives like *Halo 3,* because there is a far greater potential for variety in the sequences of events that are strung together by the players.

World of Warcraft (Blizzard Entertainment, 2004) contains no main story line, but relies instead on numerous side quests and the dense background lore that has been developed for it. These side quests can either be one-off or sequentially organized, in which case players have to complete a given side quest before moving on to the next one. The variety of activities players can engage in, the presence of other player-controlled characters and social networks, and the richness of the world's details creates the potential for a complex alterbiography that is supported by *World of Warcraft*'s lore, whether players engage with it actively or pick up traces along their travels.

Push and Pull Narrative

Ken Levine, in his 2008 Games Developers Conference lecture, outlined the difference between two forms of scripted narrative: *push narrative* and *pull narrative.* The quintessential example of push narrative is the cut scene, in which the player is a captive audience, much as when viewing a movie.

The information the game designers wish to impart is literally *pushed* at the player. In a pull narrative, on the other hand, the designers embed narrative elements in the world, such as the tape recorders in *Bioshock* (2K Games, 2008), and rely on the player to *pull* the narrative to them. Levine notes that players may easily miss out on narrative aspects in this way, but that this is acceptable because those players who are genuinely interested in the story of the game will commit to the exploration necessary to find it.

As Levine states, cut scenes are a typical form of push narrative: the story elements are imposed upon the players whether they like it or not. Action is suspended for a short period and an animated sequence delivers a segment of the storyline. In games such as *Half-Life* (Valve Software, 2004), players are able to move around or look in different directions during the course of the cut scene, but the player is still restricted to the space where the cut scene is taking place. Cut scenes are ideal devices for pushing the narrative forward and ensuring the integrity of the story delivery as intended by the designers. As Klevjer (2002) explains, when done well and at infrequent intervals, cut scenes can feel like a reward to players for completing a particular quest. This effect is particularly the case when cut scenes are placed after the completion of challenging quests, allowing players to take a break from the frantic action and thus contributing to the pacing of the game. One of the most celebrated contemporary games to make effective use of cut scenes is *Grand Theft Auto IV* (Rockstar North, 2008). For the most part, *Grand Theft Auto IV*'s scripted narrative is told through cut scenes.

Levine's pull narrative is closely related to Jenkins's (2004) notion of embedded narrative. Both advocate the use of the environment to deliver narrative in an ergodic medium. One form of pull narrative is found in objects that players have the option of picking up and that, if activated, deliver a segment of scripted narrative through written text, audio, or at times short movie sequences. These objects leave it up to the player to choose whether or not to engage with the material that is being related. According to Levine, this makes engagement with the scripted narrative of the game more powerful for those who decide to pursue it. *Bioshock*'s city of Rapture is strewn with tape recorders that contain brief audio vignettes related to the immediate environment. *The Elder Scrolls IV: Oblivion* (Bethesda Softworks, 2006) employs a similar strategy to deliver the game world's history, geography, and important characters through a range of books that can be found, purchased, or even stolen from various locations.

Levine's (2008) concept of the pull narrative importantly combines the scripted elements of narrative with the individual player's cognitive effort to form a coherent whole. As Levine's concept demonstrates, the notion of players constructing their own narrative is becoming more popular within the design community. Alterbiography similarly acknowledges the importance of the experiential dimension of game narrative in analyzing the narrative aspects of a game.

Alterbiography

Alterbiography is the ongoing narrative generated during interaction with a game environment. It is neither solely a formal property of the game nor a property of the player's free-roaming imagination. Our challenge is to account for this form of narrative generation without broadening its scope to game experience in general. Alterbiography is a cyclical process afforded by the representational, mechanical, and medium-specific qualities of a game, and actuated in the mind of the player.

So far we have established a need to differentiate between scripted narrative—the narrative content and structures explicitly written into the game by the designers—and alterbiography—the narrative content that is generated during gameplay. Aspects of the latter, experiential dimension have been discussed by a number of game theorists who have attempted to forward a positive account of narrative in games (Pearce, 2004; Jenkins, 2004; Salen and Zimmerman, 2003). These approaches make important contributions, but tend to conflate narrative generated through gameplay with game experience in general.

Like all narratives, alterbiographies focus on a character or group of characters. This concept of *focalization* was outlined by Gérard Genette (1980) in his *Narrative Discourse*. Genette describes focalization as a "restriction of field" (74), in that information about the fictional world is filtered through the perspective of the narrator. Although games often do not include a narrator, they do limit the player's perspective on the game world based on the kinds of entities they control and the format this control scheme takes. Alterbiographies can therefore focus on us, the players, in the game world, on an entity we control, or on a collective group of entities we are in charge of. The subject of an alterbiography can be any-thing from a unit of marines marooned on an uncharted island to a teen-ager's experiences at a new high school. In the case of avatar control games, the player can have an external or internal disposition to the character in question. In other words, the generation of alterbiography can feature the character as a separate entity controlled by the player or can be considered as being about the player in the world.

Different focalizations in game environments can be seen in the alter-biographies of miniatures, entities, and self. *Alterbiographies of miniatures* describe situations in game environments in which the players can control several entities at once, such as in real-time strategy games (RTSs) like *Age of Empires* (Ensemble Studios, 1997) or *The Sims* (Maxis Software, 2000), or can control a collective that is not individually simulated and represented, as in *SimCity* (Maxis Software, 1989) or the campaign mode in *Medieval II: Total War* (Creative Assembly, 2006). Certain games, like *Medieval II: Total War*, operate on multiple levels: players can issue orders to manage their cities, taxes, or diplomacy, and to perform military maneuvers. But they also control the units making up their armies in battle. If players wish, they can also participate in the battle from the point of view of the general and thus shift into the alterbiography of an entity.

The *alterbiography of an entity* describes stories relating to a single entity, which the player controls. It is differentiated from the alterbiography of self mainly by the player's disposition, although third-person games, such as *Max Payne* (Rockstar Toronto, 2001) or *Fahrenheit* (Quantic Dream, 2005), more commonly evoke this form of alterbiography.

Figure 7.1
Alterbiography focalization in game environments.

The *alterbiography of self* is most commonly evoked in first-person games like *Fallout 3* (Bethesda Game Studios, 2008) or *Mount and Blade* (Tale Worlds, 2008), where players interpret the events happening in the game as happening to *them* specifically, rather than to an external character.

In all of these modes, but especially the last two, it is always the disposition of the player that matters in determining an alterbiography's focalization. A player might be in charge of a whole football team but have an alterbiography centered on her role as the team's manager. Similarly, in a game of *Medieval II: Total War*, a player's alterbiography might be focused on the Holy Roman Empire as a whole (miniatures) or on the player as the ruler of the Holy Roman Empire (self), or it might switch between various characters in different stages of the game (entity).

Synthesis

Whether we like it or not, our adventures in Cyrodill, the world of *The Elder Scrolls IV: Oblivion* (Bethesda Softworks, 2006), will always begin in the same way: we are in a prison cell without knowing exactly what we have done to merit our predicament. As it turns out, the cell we are in was supposed to be kept vacant because it contains a secret passage leading out of the castle, serving as an escape route for the nobility in times of need. In this case it is King Uriel Septim himself who is fleeing the castle after an assassination attempt. The king recognizes us from a dream and decides to trust us and let us join him and his bodyguards as they escape through the dungeons. Matters are complicated further when assassins ambush the party. The king is killed, but hands over a pendant for us to return to a trusted friend of his. Eventually, we emerge from the dungeon into the world of Cyrodill, and the game begins in earnest.

Oblivion contains a primary scripted narrative that players can choose to interact with or ignore. The backbone of the scripted narrative consists of a series of quests that the player must complete in order to advance the events of the scripted storyline. But *Oblivion* contains far more than a single, pre-scripted storyline for players to follow. There are other, minor storylines, which at times lead to further quests. These forms of scripted narrative, however, are not the only parts of the game: The world of *Oblivion* is an open environment where players can create avatars and are then free to roam the beautifully rendered landscapes, meet other (computer-controlled) characters, and generally do as they please, at least within the constraints of the game. The game world is inhabited by fantasy creatures and people who go to work, gossip, and interact with the player. There are houses filled with everyday objects that can be picked up, tucked into one's bag to be sold, thrown, or even piled up in entertaining patterns. The open-ended gameplay afforded by an open landscape rife with objects and entities to interact with is a key attraction of games like *Oblivion*. Numerous locations are also imbued with embedded narrative elements (Jenkins, 2004) that enrich the world. In other words, the game environment has been designed to be rich with story-generating potential.

An alterbiography is the active construction of an ongoing story that develops through interaction with the game world's topography, inhabitants, objects, and game rules and simulated environmental properties. Not every assemblage of sign and code is likely to be an equally inspiring source of story generation for every player. The formation of an alterbiography is thus dependent on the disposition of the individual. Where one player might find that a game environment provides a stimulating narrative, another may not:

I'd have to say that the aspects of MMOs that interest me the most are the massive worlds created, with a multitude of storylines that run parallel, concurrent, and often cross. They are worlds in which I can create a persona and live an experience beyond what I can have in the real world, where I can create my own stories. (Rheric, *World of Warcraft*)

Nothing you do is going to affect the outcome of the scripted storyline, is it? You can only make up your own story within that framework. I think it'd be better if you could actually affect the framework and change the story. (Inauro, *World of Warcraft*)

Other players might not be interested in interacting with story elements at all. The combinatorial power of alterbiography is related to the phenomenology of reading proposed by Wolfgang Iser (1991), who similarly describes the combination of the textual properties of the printed page with the internal syntheses of the reader:

The text itself, however, is neither expectation nor memory—it is the reader who must put together what his wandering viewpoint has divided up. This leads to the formation of syntheses through which connections between signs may be identified and their equivalence represented. But these syntheses are of an unusual kind. They are neither manifested in the printed text, nor produced solely by the reader's imagination, and the projections of which they consist are themselves of a dual nature: they emerge from the reader but they are also guided by signals which project themselves into him. It is extremely difficult to gauge where the signals leave off and the reader's imagination begins in the process of projection. (135)

The concept of alterbiography combines the phenomenology of literary narrative described by Iser with the matrix of elements that make up the game environments discussed in chapter 1. The strength of the narrative disposition is always dependent on the player's inclination, but obviously a game environment with more attractive *narrative props*, to borrow a term from Kendall Walton (1990), is more likely to generate an interesting alterbiography. The perspective on experienced narrative advocated here is a mental construct (Ryan, 2006) generated by the properties of the media text. This mental construct can be derived as readily from the numeric values of a character's attributes, or *stats*, as it can be from visual and auditory representations. The concept of synthesis described by Iser is the first building block in the generation of alterbiography, which can be seen as a casual assemblage of segments of syntheses.

The concept of synthesis is not being used to signify a particular formal quality or speech act marker that identifies (and opposes) the fictional with the real. Our interest here is to create a vocabulary that will facilitate discussions about narratives generated from games in their various dimensions. Synthesis represents the culmination of the effort between designer and player, writer and reader, that becomes manifest in the player's or reader's mind. This phenomenon is not unique to games but is present in any representational medium:

When a work is produced, the creative act is only an incomplete, abstract impulse; if the author existed all on his own, he could write as much as he liked but his work would never see the light of day as an object, and he would have to lay down his pen or despair. The process of writing, however, includes as a dialectic correlative the process of reading, and these two interdependent acts require two differently active people. The combined efforts of author and reader bring into being the concrete and imaginary object which is the work of the mind. Art exists only for and through other people. (Sartre, 1967, 27)

As both Sartre and Iser argue, in the case of literature the process of synthesis between an arbitrary sign and the mental image it generates is crucial for the reading process to occur. In game environments, we interact with both arbitrary and iconic signs (i.e., verbal text, images, and audio) as well as with the rules of the game, which, if we are able to interpret them meaningfully, contribute to the process of synthesis. This process is well illustrated in tabletop RPGs.[3]

In an RPG system that expresses attributes as ranging between the values of 1 to 21, a character with an appearance value of 4 will be considered rather unpleasant-looking. Although the mental image generated by these numbers will vary among players, their imaginings are grounded in the numerical value and the rule system that gives it meaning. If we also know that the same character has an intelligence value of 18, our image will be somewhat reconfigured to take this factor into account. When the player who controls this character declares that she is striding out of the tavern where she had been drinking, she generates a succession of images based on this declaration and players' image of the character. If the RPG group is using miniatures and markings on a hex sheet to represent the individuals in the party and their environment, the process of synthesis is further modulated by the representational qualities of the miniatures, the terrain, and other markings used on the game board. This episode is an example of an alterbiography generated from stringing together a series of causally related segments of synthesis of interaction with and interpretation of game rules, representation, and mental imagery. The examples presented here are trivial in terms of their narrative complexity, but they show how even the most basic of game operations generate narrative segments that contribute to the overall narrative experience.

Tabletop RPGs are a useful example of how rules contribute to the generation of alterbiography, because the material analogs they use have poor graphical representational quality. No matter how impressive the quality of the representational layer of a game environment is, however, its experience and narrative elements come together only after a considerable amount of synthesizing work on the part of the player. Most of the time, the player is not aware of the internal work being performed in this operation, as is generally the case with perception, at least until the synthesized image is readily determinable. It is only when the qualities of the representation are ambiguous that our synthesizing efforts become apparent (Sartre, 1995).

As described in the RPG example above, interaction with the rule system of a game affords the generation of alterbiography. The same is true in virtual game environments, where a character's attributes are often expressed in numerical values. These values manifest themselves during the course of the game in various ways. A strong character can carry more items before getting tired; a charismatic character might have access to a greater number of dialog options. But there are other imaginings based on the rules which are not necessarily upheld by the system. The visual representation of the character in the form of the avatar rarely expresses these numerical values, for instance. In *The Elder Scrolls IV: Oblivion* (Bethesda Softworks, 2006), my character may have a rotund body, which remains the same as his strength, endurance, speed, and agility increase. The graphical representation does not change to reflect the change in his characteristics. Despite this, I am still able to imagine my character's body changing and developing over time. The statistical values of a character are internally synthesized with the avatar's graphical representation into a composite image in the player's mind, if the player cares at all. This is not necessarily the case in every situation, but, when narrative is generated through interaction with rules, the resultant ongoing narrative is a combination of rules, representation, and imagination.

The ways in which the overt aspects of a game system's mechanics generate alterbiographies are not limited to the attributes in Computer RPGs (CRPGs), but occur, to some degree, in the majority of games. Let us consider a completely different genre: sports games. *FIFA 2008*'s (EA Sports, 2007) management mode includes newspaper clippings that reflect aspects of the history of the team being managed and the career of the manager. After certain matches, the world surrounding the manager, which is merely alluded to and not in any way simulated, is brought to light through simple scenarios that the player has to respond to. These situations can range from the players' arguments about music in the locker room to media requests for appearances by the manager. The player's responses to these situations yield a variation in fan support, team morale, money, or the board of directors' opinion of the manager. Although not simulated or represented in any way other than textually, the scenarios stimulate the player's imagination and add to the ongoing alterbiography of the player's career as a manager. This also happens when the game system is not transparent to the player (which it rarely is), and thus interpretations of certain outcomes, like a team member's performance not living up to his numerical skills, might be attributed to the in-game entity rather than to the ineptitude of the controlling player. If Ronaldinho has three matches in which he is completely ineffective, it makes more sense for the coherence of the game world to blame the poor performance on him rather than the controlling player, since the controlling player's role in the game world is that of a manager, not a sideline puppeteer.

At other times there are graphical or audio representations in a game that have no influence on actual play. The chosen race of a character in *Oblivion* modifies her starting attributes and physical appearance, but the other characters in the game will not treat her any differently based on race alone. Despite this, if our character is of the Redguard race, we might choose to interpret a negative interaction with a Breton character as a sign of racial snobbery and react to it with anger. We might react this way because we are inclined to follow our imaginative input into the alterbiography or because our lack of knowledge about the game system prompts us to assume that this racial issue is built in when it is not. Whatever the case, our alterbiography can also be shaped by representations that are not supported by the simulation.

Alterbiography occurs in the micro phase of narrative involvement. Like all other dimensions of involvement, narrative involvement and thus alterbiography can be focused on explicitly, occupying the player's conscious attention. Players might focus on this aspect to make decisions in the game that are congruent with the already accumulated alterbiography or to consciously shape the future of the character in the game world.

The generation of alterbiography can also accumulate without conscious attention being directed toward it. Here, the existence of an alterbiography does not make itself apparent until it comes into question in one of the ways mentioned above. Narrative involvement in the micro phase is highly dependent on the formation and interpretation of the alterbiography, as it can become an important overlay which players use to interpret their actions and other events in the world.

The Relationship between Scripted Narrative and Alterbiography

The promise of an interesting scripted narrative can attract players to the game in and of itself. This attraction can vary from the general appeal of a particular setting and genre to a specific expectation of an intriguing story that players can participate in. If the alterbiography generated during the micro phase meshes well with the scripted narrative, players will tend to care enough about the game world, its events, and its inhabitants to want to return to the game in order to find out more about them and to see where the scripted narrative will lead. An engaging scripted narrative has the ability to keep players wondering what might happen next, not only when they are playing, but also between sessions. Although this can occur with digital games, literature and film tend to achieve this effect more efficiently than games. What can be particularly powerful in games, however, is that players can wonder not only about what is going to happen next but also about what will happen when they make certain choices.

In the micro phase, scripted narrative is most immediately visible when players invest their attention in events which have happened in the world prior to their arrival and which are not related to the entity or miniatures they are controlling. This emphasis on the past occurs because all elements of the scripted narrative relating to the present are absorbed into the alter-biography. What goes on *now* for the player becomes part of his or her own story. Thus, alterbiography integrates elements of scripted narrative that relate to the entity or miniatures that are its focus.

Although a distinction can be made between scripted narrative and alterbiography, they are not mutually exclusive categories. The alterbiog-raphy includes elements of scripted narrative that the player focuses on. Scripted narrative thus interacts with and molds the production of the player's alterbiography. The two intertwine when players absorb pre-scripted events in the world into their alterbiographies. In *Bioshock* (2K Games, 2008), for example, the player's encounter with Dr. Steinman is

Figure 7.2
Episodes in the life of Muun in *World of Warcraft* (Blizzard Entertainment, 2004).

Figure 7.3
Dr. Steinman in *Bioshock* (2K Games, 2008).

foreshadowed through a series of audio recordings by and about Steinman
that are picked up along the way to his clinic, as well as various references
to the demented plastic surgeon through posters, portraits, and other representational elements. The player is thus primed by these insights into
Dr. Steinman's aesthetic ideals and is aware of his degeneration into
madness. Players can ignore these clues about Steinman, but if they engage
with them, the initial encounter with the doctor is given a particular
context. When the player finally walks into Steinman's surgery to find him
operating on his dead assistant, the scripted narrative becomes part of the
alterbiography of the player. If the player cares about the scripted narrative
enough to absorb it into his alterbiography, narrative involvement is
deepened.

Narrative involvement thus concerns the player's engagement with
story elements that have been written into a game as well as those that
emerge from the player's interaction with the game. In this chapter I have
introduced the terms *scripted narrative* and *alterbiography* to distinguish
between the preordained story elements and the story generated by the
player's moment-to-moment actions within the game environment.

8 Affective Involvement

Macro Phase: Escapism and Affect

One reason for the intensely absorbing nature of digital games is the potential they have to affect players emotionally. Although other media also achieve this, an important difference with digital games is the way they place the player in a cybernetic feedback loop between human mind and machine. The player's active input creates the potential for a more intense emotional experience, whether satisfying or frustrating, than non-ergodic media provide. The cognitive, emotional, and kinesthetic feedback loop that is formed between the game and the player makes digital games a particularly powerful medium for affecting players' moods and emotional states (Bryant and Davies, 2006; Grodal, 2000).

For those suffering from a lack of excitement, games offer an immediate source of emotional arousal. Conversely, for those whose work or personal lives are too hectic, the games' compelling nature makes them ideal for shifting their attention to a performative domain that suits the players' needs: to vent frustration through intense action, to become absorbed in the cognitive challenge of a strategy game, or to stroll at leisure through an aesthetically appealing landscape. As will be discussed below, the appeal of beautifully rendered environments can be particularly powerful when contrasted with less attractive everyday surroundings.

In the field of media psychology, *excitatory homeostasis* refers to "the tendency of individuals to choose entertainment to achieve an optimal level of arousal" (Bryant and Davies, 2006, 183). That is, understimulated persons tend to choose media content that is arousing, while overstimulated persons tend to choose calmer media content. Games offer a variety of means to affect mood through interaction and allow players to tweak game settings to bring about the desired affective change. Though MMOGs are limited in the players' ability to change difficulty levels and other game settings, they make up for this by providing a wide variety of activities that can often suit different emotional needs.

Game design, like other forms of textual production, is imbued with the rhetorical strategies of affect. But unlike non-ergodic media, in games this rhetorical power is emphasized by the conjunction of textual interpretation and the performed practice of playing. The recursive input/output process inherent to game interaction has the potential to deliver powerfully affective experiences. Designers aim to capitalize on these affective qualities by selling a packaged experience that meets the expectations of buyers while engaging the range of emotions the game aims to arouse.

The compelling and mood-affecting qualities of games are often associated with the concept of escapism. Labeling an activity *escapist* carries with it at least a sense of triviality, and often more seriously derogatory connotations. Before discussing the difference between the attraction of an absorbing activity that is perceived as a pleasurable break from the everyday, and the aesthetic attraction of an environment or world which has what Tolkien (1983) called "the inner consistency of reality" (140), we must first look at the treatment and value of the term *escapism*.

The notion of escapism implies a shift from one environment or emotional state to another one that is perceived as being more favorable. Leaving a rough neighborhood to avoid the threat of danger, migrating from a war-torn country to a more stable one, or fleeing an armed assailant to safeguard one's life are all examples of escape from undesirable or downright dangerous situations. The imperative here is an attempt to move permanently away, toward a more desirable state of affairs.

Escapism similarly presupposes movement to a more desirable place or situation but, unlike the one-way connotations of *escape*, *escapism* also implies an eventual return to the point of departure. The escapist is grounded in a location or social context that must be returned to. The escapee, on the other hand, aims for a permanent, or at least long-term, change of situation. What both have in common is the striving for a better state of being. For the escapee, positive change comes from a wholesale shift, while the escapist hopes for improvement of the original situation upon return, or at least a temporary lightening of current burdens. There is no guarantee that this change will be a positive one, or that any change will even occur, but an expectation for improvement is often there.

Escapism also plays an important part in breaking away from stagnation, and so it is often a favored antidote to boredom. Games are particularly useful in this context, as they offer a variety of experiences that not only alleviate boredom but also have the potential to emotionally affect players. These emotions might not all be perceived as positive: frustration stemming from a lack of cooperation with teammates or a defeat suffered against a competing player or team is a common experience in on-line gaming, for example. This tendency to involve the emotions often serves to engage players with other aspects of the game, such as improving their skills (kinesthetic involvement), rethinking their overall strategies (ludic involvement), or working better together (shared involvement). Importantly, both positively and negatively perceived emotions stimulated by games may be better than experiencing boredom. Indeed, negative affect can, in certain instances, heighten other dimensions of involvement.

Escapism is intimately related to the uniquely human faculty of imagination. Mental imagery allows us to retain in consciousness past experiences and potential future actions along with awareness of the present moment. Our ability to imagine also inspires our emotions and passions, playing an important motivational role in our lives. Imagination allows for the possibility of mentally detaching ourselves temporarily from the present moment, which is a defining element of escapism. Many forms of escapism are, in fact, ways of engaging our imaginative faculties. The geographer and philosopher Yi-Fu Tuan (1998) links escapism with imagination and views both as defining elements of culture and humanity:

Culture is more closely linked to the human tendency not to face facts, our ability to escape by one means or another, than we are accustomed to believe. Indeed, I should like to add another definition of what it is to be human to the many that already exist: A human being is an animal who is congenitally indisposed to accept reality as it is. Humans not only submit and adapt, as all animals do; they transform in accordance with a preconceived plan. That is, before transforming, they do something extraordinary, namely "see" what is not there. Seeing what is not there lies at the foundation of all human culture. (5–6)

As Tuan argues, one form of the real is that which makes itself felt on the individual, either as a natural phenomenon or a socially created one. The real is familiar and comforting, what Tuan (1998) calls "local pattern or order" (7). The legibility of this form of the real makes it comforting because it allows for decreased contingency. Tuan importantly emphasizes that both the natural and the designed belong to the real. A determinant of the real is therefore interpretability: "What one escapes to is culture—not culture that has become daily life, not culture as dense and inchoate environment and way of coping, but culture that exhibits lucidity, a quality that often comes out of a process of simplification" (Tuan, 1998, 23).

Tuan's argument follows postmodern thinking in framing reality as always relative to context and interpretation. If reality is relative, one may reasonably argue that a phenomenon defined by the avoidance of reality is also relative to context. Once the universality of the real is undermined, so is the universality of its avoidance. If escapism is relative to context, it stops making sense to label any specific type of activity or artifact as being, in itself, escapist. An activity can only ever be seen as escapist from the particular context of a particular individual. A baker may stand in his kitchen in the early hours of the morning imagining he is a rock star, singing into a dough microphone, while elsewhere in the city a musician, exhausted after a concert, may ignore the crush of fans around him to dream of the simple pleasures of waking early and making bread. In context, both can be seen as forms of escapism, yet neither is essentially so.

Some activities described as escapist have the particular quality of occurring within aesthetically pleasing environments. One of the important ways in which contemporary game environments affect players' moods is through the representational qualities of their environments, which enable a sense of habitable space that can be very appealing to navigate. This aspect of the attraction of game worlds was mentioned regularly by my research participants in a wide variety of contexts, and was well characterized by Oriel:

When I am alone ... it's when I am relaxed and have the sounds on. I like to pretend I am actually there. ... Like sunset time at Menethil Harbor while waiting on the boat to show up. Flying through Gadgetzan at night over the ocean and looking at the stars. Listening to the crunching snow under my feet in Winterspring. Hearing crickets chirp at night when going through the forest. ... Oh, and the red and gold falling leaves in Azshara;[1] so pretty. It's peaceful. I want to live there. (Oriel, *World of Warcraft*)

Oriel's desire to imagine that she is there, inside the virtual world, is both the product of the aesthetic beauty of the game world and also Oriel's absorption into her consciousness of Azeroth as a habitable place. Tolkien (1983) has remarked on this quality of habitability and sees it as an inherent feature of textual creations he calls "secondary worlds":

That state of mind has been called "willing suspension of disbelief." But this does not seem to me a good description of what happens. ... He [the author] makes a Secondary World, which your mind can enter. Inside it, what he relates is "true": it accords with the laws of that world. You therefore believe it, while you are, as it were, inside. The moment disbelief arises, the spell is broken; the magic, or rather art, has failed. You are then out in the Primary World again, looking at the little abortive Secondary World from outside. (132)

Responding to critics of such creations and their creators, Tolkien argues that the faculty that creates and allows readers to enter such worlds is a sign of "clarity of reason" (142). Like Evans (2001) and Tuan (1998), he rejects academics who view such activity as negatively escapist:

I do not accept the tone of scorn or pity with which "Escape" is now so often used: a tone for which the uses of the word outside literary criticism give no warrant at all. In what the mis-users of Escape are fond of calling Real Life, Escape is evidently as a rule very practical, and may even be heroic. In real life it is difficult to blame it, unless it fails; in criticism it would seem to be the worse the better it succeeds. Evidently we are faced by a misuse of words, and also by a confusion of thought. Why would a man be scorned, if, finding himself in prison, he tried to get out and go home? Or if, when he cannot do so, he thinks and talks about other topics than jailers and prison-walls? The world outside has not become less real because the prisoner cannot see it. In using Escape in this way the critics have chosen the wrong word, and, what is more, they are confusing, not always by sincere error, the Escape of the Prisoner with the Flight of the Deserter. (148)

Digital game worlds continue this tradition of creating secondary worlds and strengthen the internal logic of these worlds through the use of computing power that not only represents the world but animates its agents and keeps the internal logic in check. MMOGs further extend the powers of secondary-world creation through the persistent nature of their worlds and the shared, synchronous experience this allows.

Micro Phase: Emotional Affect during Gameplay

The rhetorical strategies employed in the design of game environments are geared toward creating specific emotional responses. Nevertheless, the effects they are intended to have are not necessarily those that materialize during play. This discrepancy can be due to a variety of factors, ranging from a player's lack of interest in the particular game or genre, interruptions from other sources demanding attention, a personal interpretation of represented events that diverges from those intended by designers, or, quite simply, ineffective design. Although designers cannot dictate absolutely the specific effects of their creations, the design choices made will tend to encourage a particular kind of reaction or emotional response from the players.

A significant portion of this rhetorical power can be associated with the mode of representation of the particular medium. A number of theorists and designers have argued that graphical power is not what makes games compelling. Although it is true that the quality and beauty of visual representation does not by itself make a compelling game, the evocative power of graphics and sound should not be discounted:

I think my favorite area would have to be Duskwood Forest. Purely because it's just so Mirkwood (from *The Hobbit*). I like Silverpine Forest and the Undead starting zones around Brill for the same reason. Very atmospheric because of the lighting, sound, good monster types and placement. It's all thematically consistent. The things you'd expect to find in dark gloomy fantasy forests are all there: giant spiders, werewolves, undead. (Inauro, *World of Warcraft*)

Graphics are often the first aspect of a digital game to capture the player's attention, both when shopping for a new game and upon first playing it. It is no coincidence that major game-reviewing sites, along with the game publishers themselves, include links to screenshots and videos of gameplay. Attractive graphics often lure players toward games they would not have otherwise considered. Whether the quality of gameplay keeps them interested is another matter altogether, but for the gameplay to be trialed, some initial hook needs to be present, and graphics often serve that purpose. The graphical style of a game gives a good idea of the genre or genres the game draws from and the relevant setting. Screenshots splattered with gore and set in dark, abandoned locations are unlikely to appeal to someone looking for a colorful and cheerful game for his young daughter.

Some genres are more dependent on specific forms of graphical and audio representation than others. An FPS set in World War II will often strive to reproduce a sense of being in the period and will thus depend on a degree of verisimilitude. That being said, I find there is a distinction between those games that focus on reproducing the period itself, and those that aim to instead capture the sense of the era given by other media representations such as movies. Take, for example, two FPSs set in World War

II: *Medal of Honor: Allied Assault* (2015 Inc., 2002) and *Red Orchestra: Ostfront 41–45* (Tripwire Interactive, 2006). *Medal of Honor* draws its visual style strongly from contemporary Hollywood movies about World War II such as *Saving Private Ryan* (dir. Spielberg, 1998) and *Pearl Harbor* (dir. Bay, 2001). The lighting and palette employed in the game as well as the dramatic action portrayed on-screen are reminiscent of these movies, with several scenes reproduced wholesale either in cut scenes or as playable level sections. *Red Orchestra* takes an altogether different approach to the World War II setting, with its graphical style and audio effects both conveying a stronger sense of historical accuracy. In contrast to *Medal of Honor*, the interface information provided is limited to the level of the avatar's fatigue and injuries and the number of ammo clips remaining. Unlike the majority of FPSs, no crosshairs are provided, making it almost impossible to hit a target further than a meter or so away without bringing the weapon's iron sights up to one's point of view. This also causes players to slow or halt their movement. Also unlike games such as *Medal of Honor* or *Counter-Strike: Source* (Valve Software, 2004), shooting accurately while moving is next to impossible. Instead, the player must often kneel or adopt a prone stance in order to compensate for the weapon's recoil.

The majority of the infantry in *Red Orchestra* carry single-shot rifles that require the player to pull back a bolt after every shot in order to be able to shoot again. Although set in the same era, *Medal of Honor* includes a crosshair in the middle of the screen, making it possible to move and shoot accurately even with automatic weapons. The designers of *Red Orchestra* also pay close attention to accurately reproducing uniforms, weapons, and objects true to the period. Graphical style, audio effects, and physics come together to create very different affective experiences in these two ostensibly similar games set in the same era.

Designers of multiplayer shooter games like *Red Orchestra* and *Medal of Honor* have been at pains to instill a need for teamwork through the use of suppressing fire, in which one player shoots at the opponents' cover in order to give her teammates time to move through an exposed area. One of the problems with this gameplay mechanic is that the effectiveness of the suppressing fire will depend a lot on the ability of the person providing it. The point of suppressing fire is not necessarily to hit the opponent, but both to make it harder for them to shoot at one's teammates and to create a sense of apprehension. *Red Orchestra* manages this by combining graphical distortion and loud audio with environmental physics, resulting in a functional impediment and a heightened sense of urgency, if not panic, in the opposition players. When a piece of cover is shot, there is a loud noise, making it harder to register what is going on in the vicinity. If the players behind the cover that is being shot at try to lean out and return fire, their screen blurs and shakes, making it incredibly difficult to aim accurately. Returning fire is made more challenging when the piece of cover in question consists of destructible materials that generate dust and fragments when hit, such as stone walls or wooden window sills. Combining visual impediments with the already challenging task of aiming through iron sights and guns which are historically accurate in their limited accuracy creates an effective use of suppressing fire. This combination of graphical, audio, and physics techniques also tends to increase excitement through a heightened sense of urgency as the team supported by suppressing fire advances.

Aesthetics and Affect

A number of research participants discussed the evocative, mood-changing powers of the MMOGs' aesthetic beauty: "I love Night Elf Land. It's serene and pretty and the music is nice here. This temple is my favorite, it's so pretty. Whenever I don't feel like playing, I come here. I sit and watch the noob night elves come in, listen to their oooohs and aaaawwws. It's cute. I remember when I first came here. It's calming" (Aprile, *World of Warcraft*).

Because these virtual worlds depend on the extended participation of players, designers need to create places that inspire positive emotions in their inhabitants. The designers of *World of Warcraft* (Blizzard Entertainment, 2004) were very aware of the effect aesthetics have on players, and thus created attractive regions with varying palettes of tastefully blended colors and a design policy that aimed to appeal to a wide audience. There is a particular appeal to inhabiting beautiful landscapes where one can roam which allows for more involving experiences than viewing attractive images in non-ergodic media.

Some players find other forms of affective arousal appealing. The action-FPS *F.E.A.R.* (Monolith, 2005) is designed to maximize excitement by combining the captivating, fast-paced characteristics of FPSs with hair-raising effects borrowed from horror movies. Players progress through *F.E.A.R.* by

Figure 8.1
Interviewing Aprile in Teldrassil in *World of Warcraft* (Blizzard Entertainment, 2004).

following a linear plot that takes them from one environment to another. Although there is no possibility of veering from the episodic nature of level progression, the environments themselves can be explored in any way the player wishes, with specific events being triggered the first time an area is entered. The game alternates between combat situations and paranormal horror scenes which might be overcome only if players react in a specific way, and less active sequences that are meant to further the plot and, often, terrify the player.

The combat sequences are made more intense by the AI of the computer-controlled agents who duck, take cover, and collaborate to eliminate the player. If the player takes cover behind a wall and peeks out to take a quick shot, only to duck back, the AI agents will call for covering fire while one or two of them advance on the player, making it harder to eliminate the advancing agents without taking damage. In most action-FPS games, once an area has been cleared of AI agents, players do not need to worry about what is behind them, focusing instead on clearing out new areas. But in *F.E.A.R.*, AI agents frequently sneak up from behind the player and

Figure 8.2
Startling sequences in *F.E.A.R. 2: Project Origin* (Monolith, 2009).

knock her out with their rifle butts. Rather than being predetermined or
triggered by traversal of an area, these flanking attacks occur as a result of
the AI adapting to the player's behavior. The possibility of being attacked
from the rear leaves players in an alerted state, watching their backs more
carefully. This, in turn, leads to a more intense engagement in the spatial
and ludic involvement dimensions of the game.

The intense combat situations serve to increase the affective power of
the horror sequences that follow them. When the player is moving cau-
tiously down a corridor and peeking around every corner to avoid being
ambushed, hearing a noise behind her will cause the player to spin around
(often letting off a few rounds in panic). Seeing objects flying off shelves
instead of finding enemy combatants makes the player disoriented. Light
bulbs start swaying and flickering, adding to the sense of eeriness, and
when the player turns back to where she was headed initially, a little girl
suddenly appears out of the shadows, scampering away on all fours. By
first increasing emotional affect through fast-paced combat sequences and
then changing the way objects behave in the game world, the game creates
an intensified sense of shock and uncanniness.

The unpredictability of AI agents used in *F.E.A.R. 2: Project Origin* is
applied with even greater emotional effect in *Left 4 Dead 2* (Valve Corporation, 2009), which employs adaptive AI techniques to calibrate the pacing
of the game. The AI "director," as the developers call it, generates enemies,
weapons, ammunition, and other objects depending on the current situation in the game in order to heighten dramatic tension and emotional

investment on the part of the players. The audio segments, visual effects, and character voiceovers activated by the director can either help or hinder players, while adding to the tense and dark atmosphere of the game. As in *F.E.A.R. 2*, the unpredictability of *Left 4 Dead 2*'s assailants keeps players constantly on their toes. Even if an area has been cleared of zombies, the director may still send another zombie horde through a nearby wall if the players appear too relaxed.

Left 4 Dead 2 is an intensively cooperative game in which players need to stick together and protect each other to survive. Besides the unpredictable enemy locations, the game also heightens the panic through the special abilities possessed by some zombies. Players encounter standard zombies in every room and on every street corner who run toward them, slashing at them. But there are also special zombies, each with abilities that can be deadly for the individual player if unprotected by her teammates. The Hunter, for example, leaps on players and knocks them to the ground, hacking at them rapidly. If the Hunter lands on one of the players, she cannot get him off her by herself; she needs another player to shoot or knock off the zombie. This situation creates considerable panic, with players shouting to their friends to save them while trying to explain where they are; the situation tends to be further complicated by the fact that the other players are also being attacked. *Left 4 Dead 2* creates such an evocative experience because its rhetorical strategies of affect tightly combine the aesthetic and systemic dimensions. The tightly knit collaborative experience further heightens the emotional impact of the game since players need to communicate effectively while under ergodic and emotional pressure.

Games like *F.E.A.R. 2* and *Left 4 Dead 2* play on the tension created by dark spaces and impending danger. As the success of both games confirms, a large number of players enjoy the high-adrenaline situations the games generate. But, of course, this is only one form of affective involvement games afford. Players look for different sorts of affective experiences in games: the pleasure of aesthetically beautiful and peaceful places like those described by the *World of Warcraft* participants, the appeal of visual styles borrowed from other popular media described earlier by Inauro, or the exhilaration brought on by the startling effects of horror games such as *F.E.A.R. 2*. At times, players will sacrifice great gameplay for the chance to have experiences in specific settings they find appealing.

Salen and Zimmerman (2003) are among a number of game designers who have expressed doubt about the trend toward improving representation at the cost of innovations in design. This view has merit from the perspective of creating interesting game mechanics, but we must not forget that digital games not only attract players looking for interesting and cleverly designed game mechanics, they also attract large numbers of players who want to live a specific, packaged, experience, such as being a Formula One driver, a World War II sniper, the manager of a football team, or a murder investigator on the Orient Express. As Inauro illustrates, the attraction toward the aesthetic dimensions of games is often informed by expectations borne from other media. Digital games are not only game systems but, more importantly, are digitally mediated experiences that aim to satisfy the desires generated by movies, literature, or free-ranging fantasy.

The affective-involvement dimension thus encompasses various forms of emotional engagement, ranging from the calming sensation of happening upon an aesthetically soothing scene, the adrenaline rush of an on-line competitive first-person shooter, or the uncanny effect of an eerie episode in an action-horror game. This dimension accounts for the rhetorical strategies of affect that are either purposefully designed into a game or precipitated by the individual player's interpretation of in-game events and interactions with other players.

9 Ludic Involvement

Macro Phase: Goals and Rewards

In *Man, Play and Games*, Roger Caillois (1962) makes a distinction between the playful attitude, which he terms *paidia*, and the structured organization that rules impose on it, or *ludus*. *Paidia* represents "diversion, turbulence, free improvisation, and carefree gaiety" (13), which manifests itself in a form of "uncontrolled fantasy" (13). For Caillois, *paidia* represents an idealized purity of play that eludes all forms of enculturation, including language: "In general, the first manifestations of *paidia* have no name and could not have any, precisely because they are not part of any order, distinctive symbolism, or clearly differentiated life that would permit a vocabulary to consecrate their autonomy with a specific term" (29).

Paidia manifests itself in various activities not necessarily related to games. Caillois's examples encompass actions varying from an infant laughing at her rattle to a cat entangled in a ball of wool to the execution of a somersault. The human mind, however, is driven to make sense of itself and its surroundings, and so the chaotic freedom of *paidia* is short-lived. Caillois refers to this structuring of *paidia* through conventions, rules, and techniques as *ludus*: "It is complementary to and a refinement of *paidia*, which it disciplines and enriches. It provides an occasion for training and normally leads to the acquisition of a special skill, a particular mastery of the operation of one or another contraption, or the discovery of a satisfactory solution to problems of a more conventional type" (29).

Caillois describes *ludus* as a form of "gratuitous" structure that is imposed upon *paidia*. *Ludus* is gratuitous because players freely opt to enter into its realm of structured challenge. When the game involves several players, the parameters that structure the challenge are negotiated and agreed upon among them. When no other players are involved, the challenge is provided by the structure of the game itself.

Although certain kinds of digital games have paidic elements to them, they can never aspire to the ideal that Caillois describes because they are strongly influenced by their coded design. Even the most free-form activity in a virtual environment is constrained by the code which enables it. Flying around with reckless abandon over Liberty City in *Grand Theft Auto IV* (Rockstar North, 2008) has a degree of freedom in terms of its kinesthetic qualities, but it is important to note that these qualities are often shaped by the coded rules of the game. Actions in a game environment are therefore influenced by the ordered realm of *ludus*; the intention of the player is always limited by the conventions of a designed system.

Rules give meaning to actions and objects in the game world. As a number of authors have noted (Salen and Zimmerman, 2003; Juul, 2005), an important aspect of games is their systemic nature. The ludic aspects of games are structured by a rule system, which assigns particular values to objects, events, and entities within the system. This rule system is the machine behind the representational surface of the game and is generally intended to create a landscape of interesting choices for the players to engage with. As Thomas Malaby (2007) has argued, these rules are intended to generate interesting forms of contingency for the players to test and explore.

Game rules thus heavily inform players' interactions with the game world. As we discussed in chapter 1, in nondigital games the game can only be played if the rules are upheld by the participating players. In digital games, the machine can uphold the rule system, while at times this system is supplemented with rules agreed upon by the players. The major difference between nondigital and digital games, with regard to rules, is that in digital games the only actions allowed within the game world are those dictated by the rule system. In certain digital games, however, particularly those that take place within extended game environments, not every action the player takes has relevance within the rule system. Players can, for example, walk around an area simply to see the sights or to kill time while waiting for a friend to come on line. Although players are interacting with the properties of the environment, such as the speed of movement, density of obstacles, and so on, these are not being considered as part of the rule system. Going back to the distinction we made in chapter 1 between simulated, environmental properties and the coded rules of a game, these play very different roles in organizing players' experience within the game world.

The rule system of a game frames the representational layer with the values dictated by the rules. This is a crucial difference between the significance of signs delivered through ergodic versus non-ergodic media. A national flag seen in a movie carries with it the usual connotations of national identity and patriotism (among other things) that flags denote. In contrast, in a capture-the-flag game of a multiplayer FPS like *Call of Duty IV* (Infinity Ward, 2007), the flag becomes less of a national symbol and more of a ludic objective which organizes the game mechanics of that particular game mode. The flag carries a host of meanings that can only be understood through interaction with the game rules. As an example, imagine that the score of a capture the flag game is tied, with thirty seconds to go before the match ends. Our team's flag was being carried to the opposing team's base, but we manage to shoot down the flag carrier. The flag is now on the ground a few meters away from the opponent's base. If the opposing team grabs it and runs it to the base, we lose the match. Our entire team realizes with alarm that the flag must be defended and retrieved or there is a good chance that the enemy will capture it and win the game. The flag thus is not interpreted as a symbol of national identity, but becomes the locus of a competitive struggle between two factions and a source of added excitement for players on both teams.

Although rule systems play an important part in drawing players into a game and keeping their attention during the moment of gameplay, rules are not often the focus of conscious attention. Rather, the rule system manifests itself experientially in the form of decision making and the pursuit of personal and game-defined goals. Research participants frequently alluded to their involvement with the rule system of the game, but they expressed this engagement primarily in terms of goals rather than the rule systems themselves. The rest of this chapter will therefore address ludic involvement primarily through interaction with hierarchies of goals and the decision-making process that surrounds such interactions.

In the macro phase, the dimension of ludic involvement revolves around the motivation for game engagement provided by the pursuit of goals and the attainment of associated rewards. During gameplay, in the micro phase, ludic involvement concerns decision making and plan formulation that enables players to work their way through a hierarchy of goals either self-assigned, set by the game, or set by other players.

Goals

Choices in games tend to be made in relation to the pursuit of goals set by the game or by the players themselves. Pursuing a goal can be an end in itself—an autotelic experience yielding satisfaction within a rule-based system of meaning—or it can be fueled by the desire of attaining a reward. We get frustrated when we fail to achieve these goals and elated when we succeed. But often the goals that are set for us, or those that we set for ourselves in everyday life, do not guarantee a reward or satisfaction (McIntosh, 1996). We may want recognition at our workplace in terms of a promotion, but factors outside our control may conspire against our reaching this goal. We can act upon known courses of action that might improve our chances to succeed, but the final outcome may be beyond our control or abilities. Games constitute systems that can be interpreted and modulated to suit the skills of the participants (Egenfeldt-Nielsen, Smith, and Tosca, 2008; Juul, 2005), thus setting up opportunities for attaining goals in circumstances which make it easier to eliminate external, unknown factors. In other words, games appeal to our tendency to organize the complexity of our lives into hierarchies of goals by creating systems in which these goals are explicitly stated and specifically designed to be attainable:

If goals have been accepted as a prominent guiding principle of the human psyche, it is even more relevant to acknowledge their prominence in relation to games. ... [G]ames in fact distil the abstract life goals that we struggle with on a daily basis into highly concrete, temporally and spatially circumscribed events, often spiced with fictional aspects that lift them above everyday struggles. (Järvinen, 2008, 131)

As a number of theorists have pointed out (Juul, 2005; Salen and Zimmerman, 2003; Järvinen, 2008; Björk and Holopainen, 2005), goals play a central role in games: they define and give value to certain actions within a game's rule system, whether the system is upheld by a machine or negotiated by players. But goals are not only present when specifically implemented in rule systems; open virtual environments also support the setting of goals by individual players. *Grand Theft Auto IV* (Rockstar North, 2008) can best be described as a virtual environment which has a series of minigames embedded in it and whose scripted narrative has a linear structure of progression with strict ludic properties. To advance the scripted narrative, players must complete missions that usually involve a number of linked goals which can be achieved only by adhering to a set of relatively strict parameters. At the same time, the game's environment supports other, player-defined goals. We may wish to drive around Liberty City looking for a nicer looking car to steal, for example. Such an objective could be seen as a form of what Roger Schank and Robert Abelson (1977) describe as an "instrumental goal" (74), a goal that acts as a prerequisite to another goal; or it could be something we do for its own sake. We might want to drive around the city in a more attractive car than the one we have simply for the sake of it, without any higher-order goal in mind. On the other hand, our desire for a nicer looking car might be related to our desire to impress our virtual date for the evening.

This example highlights the importance of distinguishing between goals that are set and validated by the coded game system and goals created by the player. *Grand Theft Auto IV* also allows players to interact with other players in their city. They can play one of the various minigames, such as pool or bowling, roam freely through the city, or decide to set up games within the environment. We could, for example, arrange to race from the entrance of the hospital and around the block for three laps. In doing so, we would have created our own game within the environment with a goal, or set of goals, negotiated and upheld by the players.

Goals can thus be determined by the game system, set by the individual player, or negotiated by a community of players, in the case of multiplayer games. Personal goals can be separate from those established by the game, and thus their scope depends on the degree of open-ended play allowed by the game system. In a game of *Space Invaders* (Taito, 1978), for example, we might set ourselves the goal of clearing the first five levels without losing a single life. Since the range of activity that the *Space Invaders* game system allows is rather limited, the personal goals that a player can set are equally limited. In contrast, more open-ended sandbox games like *The Elder Scrolls IV: Oblivion* (Bethesda Softworks, 2006) or *Fallout 3* (Bethesda Game Studios, 2008). These games provide players with an open-ended environment in which they can roam freely without the restrictions to actions and navigation that more linear games impose. Sandbox games allow players to decide their course of action in the game environment without penalizing them for not following the set storyline. In *Oblivion*, we can decide not to save the world from the hordes of demons threatening to overrun it and instead make a career of robbing people's houses for our own gain. In this case, our long-term goal could be simply to accumulate as much wealth as possible or to gain a reputation as a notorious bandit. Short-term goals would possibly include robbing rich people's houses, practicing useful skills like lock picking, and finding an adequate place to use as a base of operations. Games like *Oblivion* are attractive precisely because they are designed to react to our actions, and thus they become more meaningful to us. The more criminal activity we engage in, the more notorious our actions will become and the more likely it will be for the inhabitants of the world to react negatively to us and for the authorities to attempt to arrest us. Thus, our personal goals are validated by the game's rule system, giving our actions positive and negative repercussions that are upheld by the system.

MMOGs offer even greater scope for determining personal goals, although the goals are still limited by the coded mechanics of the game. As a first-level character in *World of Warcraft* (Blizzard Entertainment, 2004), for instance, we cannot immediately set out to explore the world, as each region has aggressive mobs that will attack and kill our weak character. Nevertheless, MMOGs allow for great scope in terms of self-created goals due to the size of their worlds, the various styles of play accommodated, and, most importantly, the presence of persistent communities of other players. Participants in my research expressed their attraction toward having the opportunity to follow personal goals in MMOGs. Ananke gave an account of her engagement with the pursuit of a personal, self-created goal:

I once swam from Darkshore. Took the better part of an afternoon and I did it just because I wanted to and it was one of the few times that I didn't mind something taking so long. It wasn't something that everyone has done. It was an adventure. I didn't know if I could or not. A nice challenge and not something that the game mechanics guided me through. (Ananke, *World of Warcraft*)

For players who are more oriented toward maximizing their character's progress in the game world (what is often called *the numbers game*), Ananke's long-distance swim might seem pointless because it did not yield specific game rewards, but for Ananke it was an enjoyable exploration of the limits of traversing game space. This is an example of how dimensions of involvement weave together in practice. The satisfaction of goal completion is made possible by the sense of agency created by allowing players to decide to do something and then to actually do it. This, in turn, is enhanced by the exploration of the game space and game system, while the chain of events that resulted in the story of the journey contributed to Ananke's overall engagement with the game world. Players derive a sense of satisfaction from feeling that they have the liberty to create and strive toward their own goals. The players' ability to work toward their own goals is thus important not only in terms of satisfaction derived from reaching the goal, but also because it gives them a sense of freedom of action and control over their experience in the world, as was emphasized by Sunniva: "It allows you to control more of your own gaming experience and thinking out your own quests, makes the game more engaging, I think, 'cause you need to plan more and think more around what you're doing" (Sunniva, *Planetside*).

The Structure of Goals

Within cognitive psychology, goals tend to be viewed as hierarchically structured (Bell and Huang, 1997; Locke and Latham, 1990; McIntosh, 1996; Rosenbloom and Newell, 1983). The exact nature of every level in the hierarchy varies from theorist to theorist, but the one distinction that the majority of researchers in the field agree on is that between higher- and lower-order goals. These are sometimes called "primary goals" and "subgoals," respectively. The more immediately attainable subgoals tend to be pursued in order to accomplish primary goals (Houser-Marko, 2007).

In his doctoral thesis, Aki Järvinen (2008) applies the distinction between primary goals and subgoals to games, arguing that games contain both lower- and higher-order goals and that lower-order goals tend to be followed in an instrumental capacity in the pursuit of higher-order goals. Staffan Björk and Jussi Holopainen (2005) similarly identify hierarchy of goals as a recurring design pattern in games. They state that when players are aware of the way in which the lower-order goals affect the higher-order goals, the former have a greater impact on player satisfaction. When playing *Fallout 3* (Bethesda Game Studios, 2008), for example, our higher-order goal might be to find our fugitive father. To find him, we must go to a radio station a substantial distance away. The intervening area is fraught with perilous obstacles, including rabid dogs, gang members, and bands of mutants. In order to successfully traverse it, we decide to find some way of procuring better equipment and thus start looking for some kind of human settlement that might contain a shop. Upon finding a town, Megaton, we realize that we must accumulate bottle caps, the new currency, in order to fulfill the goal of buying better equipment. Consequently, we will either have to steal items and try to sell them elsewhere, or work for the money. We decide to find a job.

The *Fallout 3* example describes a sequence of subgoals—procuring equipment, finding a town, earning currency—aimed at achieving the primary goal of finding the fugitive father. This identification of a primary goal is essential to keeping players motivated and engaged with the game, for the primary goal and the consequent chain of subgoals give meaning to players' actions and provide a motivation for their continued engagement. Of course, we might choose an alternative primary goal in *Fallout 3* to that suggested by the game system. We might decide that our character has lost all faith in humanity after being abandoned and left in the harsh reality of postapocalyptic life. As such, we might proceed to murder every living human in the game world. The game's rule system supports this sort of primary goal and allows us, once again through the structuring of a chain of subgoals, to work toward it. If, on the other hand, we notice that people we murdered reappear a few minutes later, then we will not be able to reach our personal primary goal. This would be the case in games like *Far Cry 2* (Ubisoft Montreal, 2008) which automatically respawn downed enemies infinitely. Different rule systems thus offer varying degrees of support for the creation of personal primary goals.

The lack of a clearly defined primary goal in a game environment can be confusing for players, while the requirement of adhering to a system-determined primary goal can feel restrictive. The type of goal available depends on the kind of game in question and the expectations surrounding it. Few players, for instance, would expect a tennis game to offer an open world supporting a variety of exploration-related personal primary goals. On the other hand, if an MMOG fails to present this kind of open, explorable world, many players will be disappointed. Thus, the possible primary goals of a game are dependent on the affordances of the game system and the virtual environment in which it is located.

While this distinction between different levels of goals is fruitful for understanding the goal structures of games, it is also the case that the entire game activity can be seen as a subgoal of other primary goals, such as feeling better through relaxation, improving one's sense of self-worth through overcoming other players, and so on. Thus, a game's primary goal tends to be only a subgoal in the larger life of the player. Certain properties of goal hierarchies in life do not translate to the game situation, however. William McIntosh (1996) states that "a related feature of the goal hierarchy is that goals located higher up in the hierarchy are more abstract than goals located lower in the hierarchy" (56). Although this can be true of primary goals in some games, it is not the case generally, because games are designed to keep players engaged over an extended period and thus benefit from having clearly outlined goals communicated to the player at all times. The primary goal of attaining the highest level in an MMOG is one example of a common primary goal that is measured at all times by the system and displayed explicitly in order to keep players striving for their goal. As one participant puts it: "Then came the numbers ... quickly getting higher levels, higher skill points, better armor (with higher numbers), more back space to carry more and more items, making more money for bigger items ..." (Oriel, *World of Warcraft*).

In everyday life, such unambiguous feedback is rarely available. When precise goal-related feedback does exist, it is seldom as clearly structured, transparently communicated, and certain as in-game goals are. It therefore makes sense to make a clear distinction between in-game goals and life goals when considering the structure and dynamics of primary goals and subgoals.

Micro Phase: Plans, Goals, and Rewards during Gameplay

Game designer Sid Meier described games as "a set of interesting choices" (quoted in Rollings and Morris, 2000, 38). Meier noted that players' ability to formulate possible courses of action and to act upon them is a key feature of games. While, as discussed in chapter 1, the wide range of artifacts we call games vary greatly in properties and scope, they all share the need for a player to interact with them. Most actions are borne out of a planning and decision-making process. Ludic involvement in the micro phase concerns engagement with the decision-making process and the perceived consequences of such actions. As discussed in chapter 3, ergodicity is made up of cognitive activity in plan formulation and readiness to act, as well as the execution of the act itself.

Plans and Goals

The execution of actions is the actualization of a plan to attain a goal. One of the engaging aspects of digital games is the iterative process of planning, execution, feedback, and replanning. Players establish which subgoals they will work toward and formulate a plan to achieve them. As the plan is executed, players are given feedback about the likelihood of success, giving them a chance to carry on with the original plan or to modify it to suit the emerging situation. The players might not have the required resources and abilities to carry the plan through, or they might perceive they are not performing as well as they could and thus decide to restart and try again. If the plan is carried out successfully and the subgoal is met, a new subgoal is engaged, often with preference given to subgoals higher up in the goal hierarchy. Ludic involvement in the micro phase thus concerns the initiation of a process of actualizing plans through the exploration of available options and the related perceived consequences, pertaining to systemic, personal, and social goals in various parts of the hierarchy.

Planning a journey from Ratchet to Feathermoon in *World of Warcraft* (Blizzard Entertainment, 2004) is an instance of ludic involvement in the micro phase. Plans are made and amended on the spot whenever unforeseen circumstances arise, such as an unexpected skirmish between player factions in an area to be traversed. The complexity of the situations that arise in multiplayer games, particularly in the open worlds of MMOGs, cannot be wholly subsumed under the scope of game rules. Whether it involves deciding on the prices of items at the auction house in *World of Warcraft*, planning a defensive strategy in *Counter-Strike: Source* (Valve Software, 2004), or deciding how many armies to devote to the siege of Aragon in *Medieval II: Total War* (Creative Assembly, 2006), ludic involvement in the micro phase accounts for the moment-by-moment assessments that accompany actions in a landscape of possibilities.

To give another example, customizing a creature in *Spore* (Maxis Software, 2008) to fit a specific mental image of what the creature should look like (personal goal) and building a functionally powerful creature that will aggressively eliminate opponents (systemic goal) both concern ludic involvement since both processes are made up of a series of choices toward achieving a particular goal. The activity need not be agonistic in nature; to be considered ludic involvement, it needs to be geared toward a specific goal.

Figure 9.1
Building a creature in *Spore* (Maxis Software, 2008).

Strategies, Tactics, and Microtactics

The distinction between primary goals and subgoals entails a commensurate distinction between the plans relating to the pursuit of each. The plans made in pursuit of primary goals will be referred to as *strategies*, while the plans made in pursuit of subgoals will be called *tactics*. Of course this is a general distinction, as there are various levels of subgoals in a hierarchy. With regard to planning during moment-to-moment actions, a further level of specificity is useful. Although all immediate actions ultimately contribute to the completion of a subgoal, it also makes sense to distinguish the plans made during the moment of action; we will refer to these plans as *microtactics*.

Consider fencing. Our primary goal may be to become a better fencer and ultimately to be ranked in the top three in the county. In order to reach this goal, we must place in top spots in a number of ranked tournaments. Thus, we now have a concrete goal to win or place well in an upcoming tournament. During the tournament, our subgoals can be broken down to winning individual matches. The strategies we employ require a training regime, the careful choosing of which tournaments to attend, and so on. The tactics, on the other hand, relate to winning individual matches. At a lower level still, within the individual match, are the on-the-spot decisions we make to attain the immediate goal, which is to win the current point. A series of microtactics are employed in pursuit of this immediate goal. The notion of microtactic delineates a further level of specificity when describing plans relating to a rule system. In using the term, I do not intend to rigidly label actions, but to express the hierarchical relationship between two orders of planned action. Going back to our fencing example, the tactic for the individual match might be stated: "Disrupt the opponent's attacks while retreating, and use counterattacks to score safe points." The microtactic would be: "He is out of balance, with the foil moving clockwise; lunge with a counterclockwise evasion of the blade."

These levels of specificity of planning and execution are useful only for analyzing activities in which there is a short temporal distance between plan formulation, execution, feedback, and evaluation. In a board game, it would not be particularly useful to talk of microtactics, because players have time to think about their tactics and strategies. In the majority of game environments where players need to act, the concept of the microtactic identifies decisions to immediately perform certain actions instead of others.

One personal primary goal we might have in the Ultimate Team mode of *FIFA 09* (EA Sports, 2008) is to create a team consisting solely of Italian players and to win an on-line cup against other human players. There is a long sequence of subgoals we must follow before we can even start recruiting players. We need to make money by playing matches, while not losing many of them. Since each match played reduces the amount of contract time each player has and extending these contracts is expensive, we need to use players who we do not intend to keep in the long run. We then need to acquire a good team of staff, including a manager, head coach, goalkeeper trainer, physiotherapist, and fitness instructor. We need to decide which of these staff members are indispensible for our initial progress, as they all cost money and contract time. Then we need to acquire and develop our chosen players without using up their contract time. At the same time, we need to keep the morale of these players at good levels, which is hard to do if they never play matches.

This small example shows the complexity of connected subgoals and the related tactics a player employs to achieve them. The overall strategy will probably shift depending on the perceived possibility of attaining the primary goal. All these examples deal with subgoals during the management part of the game. While the key tactics we employ are in pursuit of the subgoal of winning the match, another important subgoal might be to avoid player injuries or send-offs. During the actual game, though, we focus on the subgoal of scoring and preventing the opposing team from scoring. These subgoals are executed through a sequence of microtactics such as ball passing, crossing, dribbling, tackling, and so on. The situated moment of gameplay requires us to make rapid decisions. If our right winger is running up the sideline and is faced by a defender, we have to make a quick decision: Do we perform a short pass to the right back running behind us, do we try to dribble around the defender and then cross, or do we cross the ball now and hope that the defending line does not intercept it? The individual decision is a microtactic in service of higher-order tactics that are deployed in pursuit of a series of increasingly higher-order goals.

Rewards

Aside from the satisfaction one feels in attaining a goal, games often tempt the player with specific in-game rewards. These rewards are delivered to the player incrementally as she makes her way through the structures of progression designed into the game. Rewards can vary greatly, sometimes awarded and validated by the system, while at other times having a more personal or social nature. The value of a game reward is essentially subjective and always depends on the player's wants, needs, and expectations. The same cut scene can feel like a reward to some players and an annoyance to others. Finding a rare magical hammer in an MMOG might seem like a great reward for a player who needs it (or the monetary and/or social value it might yield), while to a wealthy player who has a rack of better magical hammers, it might be a disappointment.

Rewards can improve the abilities of the player's avatar or miniatures, making it easier for them to accomplish certain tasks or giving them altogether new abilities. These forms of reward enhance the performative ability of the player. In *Planetside* (Sony Online Entertainment, 2003), attaining a new level gives the player a skill point that can be used to buy access to new armor, weapons, and vehicles. These skill points increase capabilities even further, because each new skill has the potential to enhance the utility of other skills. If, for example, we are able to put on an infiltrator suit (a suit that makes the wearer nearly invisible) and then try to hack at the door of an enemy facility to infiltrate their base, our abilities will be greatly enhanced by a new addition that allows us to operate reconnaissance aircraft. Flying at great speeds over enemy towers and then dropping onto the rooftop will make it far easier for us to hack into the tower than if we had to walk or drive up to the tower in a land vehicle, especially if battle is raging around it.

Some of these improved abilities afford players greater mobility, allowing them to navigate and learn space in a more efficient or attractive manner. These abilities are particularly desirable in environments that span great distances or in situations where spatial information is not available to players in the game's default setup. In large game worlds such as Cyrodill in *The Elder Scrolls IV: Oblivion* (Bethesda Softworks, 2006) or Middle-earth in *Lord of the Rings Online* (Turbine, 2007), getting from one point in the world to another can take hours, especially when no teleportation facilities exist or when these facilities are not accessible to the player. Thus, means of traversing space in a more expedient manner are highly desirable. In MMOGs, these usually come in the form of mounts or vehicles, depending on the game world's setting. In *Oblivion* or *Lord of the Rings Online* players can purchase horses (or ponies, if one's character is small), which greatly speed up navigation and exploration in the world.

In games where the player is not tied to an avatar but has a mobile and omniscient viewpoint, spatial rewards take the form of added visibility of the game map, rather than improved mobility. In certain real-time strategy games that implement a game mechanic known as the fog of war, the reward of added visibility is an important factor in making tactical decisions and keeping informed of the enemy's movements. Although the player's viewpoint is omniscient, areas in the map that are not in the line of sight of the structures and miniatures controlled by the player are presented as dark or are otherwise obscured. Some RTS games reward players with limited special abilities that allow them to see what is happening in a section of the obscured zone. Similarly, in games such as *Grand Theft Auto IV* (Rockstar North, 2008), space itself is used as a reward, with players gaining access to new areas for exploration by completing the game's various missions.

Of course, the very act of completing a mission or task can be a reward in itself. Certain goals are pursued due to their intrinsic value within the game system. One straightforward example of this are the points assigned to players in arcade games when they perform successful actions, ultimately building toward a high score, as in *Space Invaders* (Taito, 1978). In a related vein, some FPS games such as *Counter-Strike: Source* (Valve Software, 2004), reward players with exaggerated sound effects, such as a loud voice saying, "Monster Kill," when they are performing well during a round of combat. Another example of such rewards is the achievements system common to contemporary console games. Upon the completion of a predefined goal, an achievement is unlocked and displayed on the player's profile. This functions as a form of social reward, where the satisfaction of completing the goal takes the form of bragging rights given to that player in their gaming community. Often, players only care about "unlockable" achievements such as a sense of personal satisfaction in accomplishing the relevant feat in the game.

As research participants stated, seeing their scores increase and level meters filling up is at times enough of a reward in itself. This is sometimes referred to as *the numbers game* because players can see their quantified accomplishments accumulate, often expressed in numerical or bar graph format. This format is an attractive aspect of level-based RPGs and MMOGs, where missions or quests yield a certain number of experience points (XP) which go toward character advancement and other forms of monetary or item rewards. Each quest therefore constitutes an attainable subgoal that contributes toward the higher-order goal of developing a character through increasing levels, each level of which is a goal toward which players tend to work. While the reward of reaching a new level often yields added abilities, it can also be a purely ludic reward that is pleasurable in itself, particularly when the player has had her eye on her slowly increasing level meter. There is something alluring about seeing that bar fill up until the familiar level-up sound is heard, only to start the process again. Participants described how they would often continue playing longer than they had intended in order to reach a new level, to give the game session a sense of closure and to log off with the satisfaction of knowing that a concrete game goal was reached. Sometimes they would also pursue avenues of action that they found tedious in order to reach their goal, the most common form of which is known as *grinding*. Grinding is the act of repeatedly killing unchallenging creatures with the sole aim of gaining experience points as efficiently as possible. Although players admit that the activity is generally anything but enjoyable, the lure of reaching the next level is so strong that it makes the tedium feel worthwhile: "[It's] pretty much the only way to level. This is so not like the other MMOs, can't count on people here for the most part. So, I do what I have to do to reach my goals: to get 60, be able to do raids/instances, obtain items I want/need" (Aprile, *World of Warcraft*).

The lure of ever-increasing statistics is not limited to MMOGs. Games which quantify the improvement of the player's character, football team, city, army, or household members (to name just a few) have the potential to hook players into the desire for higher and higher numbers. As several participants expressed, the presence of a clearly quantifiable improvement validated by the game system is a strong motivational factor in itself.

Figure 9.2
A player's character gaining a level in *Aion: The Tower of Eternity* (NC Soft, 2009).

The numbers game becomes more potent when the numerically quantified goal is compared to that of a competing player or group of competing players. These forms of competitive rewards can take a number of forms, from the satisfaction of having the most kills in a round of *Counter-Strike: Source* (Valve Software, 2004) to being the first guild to defeat a challenging boss during a raid in *World of Warcraft* (Blizzard Entertainment, 2004).

Aside from competitive rewards, multiplayer games also offer rewards that enhance collaboration. In *Lord of the Rings Online* (Turbine, 2007), for instance, the Burglar class can conceal friends in their party, allowing them to move undetected among enemy creatures. These kinds of abilities give the player an enhanced sense of utility among fellow players. The requirement of reaching a particular level in order to unlock such collaborative abilities can thus become an appealing goal for the player, since it gives her a greater ability to aid others. In class-based MMOGs, players make a choice about the kinds of goals and rewards they are going to work toward when they choose their starting class. The Priest class in *World of Warcraft* is an example of what are called *healing classes*. The role of players in such classes is to assist their companions by protecting and regenerating them when they take damage. Since the healing classes are designed to assist others, the personal, systemic, and social goals they complete tend to reward them with more efficient and interesting ways to support other players.

An altogether different means of rewarding players is through the scripted narrative, where one exists. When they complete goals set by the game, players are rewarded with preset sequences or cut scenes that forward the scripted narrative. In *Grand Theft Auto IV* (Rockstar North, 2008), for example, the completion of goals relating to the main narrative arc of the game triggers animated cut scenes that gradually tell Nico's story. This works well in *Grand Theft Auto IV*, as the animated scenes are graphically appealing and, more importantly, portray interesting characters that are given a degree of depth through the quality of the script, the voice actors, and their visual representation. In other cases, the supposed reward may become more of an annoyance, particularly if players do not have the option of skipping the sequence of scripted narrative. It can be especially tedious when a player has to endure a lengthy cut scene they care little for: not everyone enjoys watching an hour-long movie during gameplay, as happens in *Metal Gear Solid 4* (Kojima Productions, 2008). But when the scripted narrative captivates players, this form of reward can be a potent attractor to continued engagement with the game, since they will want to continue playing to discover what happens next.

Achieving or failing at goals tends to create an emotional response in players. Although they can both calm and excite the player, the majority of digital games on the market today are directed more toward stimulation than calming. The player's active participation makes the very act of achieving or failing to achieve a goal an emotionally charged experience, as the outcome is a reflection of his dexterous, cognitive, or social abilities. The very act of achieving a desirable goal, therefore, has an associated sense of satisfaction that in many ways is its own reward. Conversely, failing to achieve a goal that a player feels is within his abilities can create a sense of disappointment or frustration. On the calmer side of the affective dimension, rewards can also be tied to less ludically challenging and more harmonious experiences, such as the emotional impact resulting from the appreciation of aesthetic beauty. Coming across a beautiful view over a harbor at sunrise, as in *Grand Theft Auto IV* (Rockstar North, 2008), is a

Figure 9.3
Aesthetic beauty and goal completion in *Flower* (ThatGameCompany, 2009).

reward in its own right. A game like *Flower* (ThatGameCompany, 2009) explicitly uses aesthetic beauty to reward players: when a sequence of flowers is completed, the surroundings burst into color.

Ludic Involvement Summary

Whether we are building a powerful character or taking over the planet, ludic involvement is one of the most fundamental cornerstones of a game. It is compelling due to the landscape of cognitive challenges games are so apt at presenting and their associated rewards. Overcoming challenges and attaining goals create a sense of accomplishment and satisfaction in the player.

Ludic involvement concerns players' engagement with the choices made in the game and the repercussions of those choices. These choices can be directed toward a goal stipulated by the game, established by the player, or decided by a community of players. They can also be spur-of-the-moment decisions with no relation to an overarching goal. Without repercussions, actions lose their meaning, and thus their execution generates no excitement.

Now that we have considered each dimension of the player involvement model in turn, we move on to consider how the model addresses the four challenges discussed in chapter 2. In so doing, I will propose a solution to the question of presence/immersion that emerges from the model.

10 From Immersion to Incorporation

Game environments afford experiences that are not available through non-ergodic media. One of these experiential phenomena is the potential to metaphorically inhabit their virtual spaces not just through our imagination, but also through the cybernetic circuit between player and machine. As we saw in chapter 2, the metaphors of immersion and presence that have been used to express this phenomenon are complicated by terminological confusion and vagueness. In chapter 2, we identified four specific conceptual challenges that have created the majority of the difficulties those terms face: (1) researchers in game studies have used the term *immersion* interchangeably to mean *absorption* or *transportation*; (2) the qualities and affordances of non-ergodic media need to be taken into consideration when understanding immersion; (3) immersion is not determined solely by the qualities of the technology being used; and (4) various forms of experience that make up immersion need to be considered as located on a continuum of attentional intensity.

Along with those challenges, defining *presence* and *immersion* is further limited by describing the experiential phenomenon we are addressing in terms of a unidirectional plunge into a virtual world. As we shall see, this is not the best metaphorical entailment and, in particular, this assumption—that the external world can be *excluded* so completely as the participant is submerged *into* the virtual environment—is problematic: The player or participant is not merely a subjective consciousness being poured into the containing vessel of the game. Our awareness of the game world, much like our awareness of our everyday surroundings, is better understood as an absorption into our mind of external stimuli that are organized according to existing experiential gestalts (Lakoff and Johnson, 2003; Damasio, 2000). If we feel that we exist in the game world, it is because the metaphor of habitation it provides has a sufficient fit with the experiential gestalts that inform being in everyday life.

As the complexity and sophistication of game environments increase, the metaphors of everyday life become more easily adaptable to experiences within them. By *everyday life*, I am here referring to the composite nature of contemporary being, in its social and media-saturated cultural dimensions. The appeal of otherness that these environments promise is organized by the same structuring principles of the everyday world. This is the power of the composite phenomenon that *presence* and *immersion* allude to: a process of internalization and experiential structuring that is compelling precisely because it draws so strongly from everyday lived experience. George Lakoff and Mark Johnson (2003) emphasize this dynamic of transference between experiential gestalts as the core of their experientialist ontology:

Recurrent experience leads to the formation of categories, which are experiential gestalts with those natural dimensions. Such gestalts define coherence in our experience. We understand our experience directly when we see it as being structured coherently in terms of gestalts that have emerged directly from interaction with and in our environment. We understand experience metaphorically when we use a gestalt from one domain of experience to structure experience in another domain. (226)

As Lakoff and Johnson stress in their work, meaning results from the interaction that takes place between language and lived experience, each of which modifies the other in a process that is crucially metaphoric. Metaphor is not simply a deviation from a literal reality; it is created by, and in turn creates, our sense of reality (Richards, 1936; Lakoff and Johnson, 2003). This process of mutual validation is particularly relevant in the case of more abstract experiences, where figurative expressions facilitate the structuring necessary for the relevant experiences to be internalized.

The metaphor we should use to understand the sensation of inhabiting a virtual environment would best draw upon our knowledge of the experience of inhabiting the everyday world. I have been careful to avoid contrasting the virtual world with the real world, as is often done in discussions of virtual environments, precisely because taking this distinction for granted is misleading (Lévy, 1998). Virtual environments are an important part of our everyday life and are more productively seen as deeply interwoven with our sense of reality. A metaphor of virtual world habitation, therefore, should draw upon the experiential gestalts of everyday habitation; that is, a view of consciousness as an internally generated construct based on the organization of external stimuli according to existing experiential gestalts (Dennett, 1991; Damasio, 2000; Lakoff and Johnson, 2003).

The metaphors of immersion and presence, founded as they are on an exclusionary logic, do not enable such a perspective on the phenomenon. I therefore propose the metaphor of *incorporation* to account for the sense of virtual environment habitation on two, simultaneous levels. On the first level, the virtual environment is incorporated into the player's mind as part of her immediate surroundings, within which she can navigate and interact. Second, the player is incorporated (in the sense of embodiment) in a single, systemically upheld location in the virtual environment at any single point in time.

Incorporation thus operates on a double axis: the player incorporates (in the sense of internalizing or assimilating) the game environment into consciousness while *simultaneously* being incorporated through the avatar into that environment. The simultaneous occurrence of these two processes is a necessary condition for the experience of incorporation. Put in another way, incorporation occurs when the game world is present to the player while the player is simultaneously present, via her avatar, to the virtual environment.

We can thus conceive of incorporation as *the absorption of a virtual environment into consciousness, yielding a sense of habitation, which is supported by the systemically upheld embodiment of the player in a single location, as represented by the avatar.* This conception retains the two traditional interpretations of the term *incorporation*: incorporation as a sense of assimilation to mind, and as embodiment. While I am aware that the latter is the sense that has received more attention in the humanities and social sciences, the work to which the term is being put here treats both meanings as essential.

The player involvement model provides the foundation for this new conception to be built on. It provides a multifaceted understanding of involvement, ranging from off-line thinking and motivation to the involvement experienced during specific instances of gameplay. Incorporation can thus be described as an intensification of internalized involvement that blends a number of dimensions. It is a synthesis of movement (*kinesthetic involvement*) within a habitable domain (*spatial involvement*) along with other agents (*shared involvement*), personal and designed narratives (*narrative involvement*), aesthetic effects (*affective involvement*), and the various rules and goals of the game itself (*ludic involvement*).

Although many combinations of dimensions can result in incorporation, two particular dimensions—spatial and kinesthetic—form the cornerstone of the incorporation experience. Without them, incorporation cannot take place. Kinesthetic involvement and spatial involvement combine in the process of internalizing the game environment. Players must be capable of movement, the ability to navigate space, in order to be involved with, and later internalize, the spatial dimensions of the game environment. Until players learn to move in the world, they cannot engage with its spatial dimensions. A player of *Counter-Strike: Source* (Valve Software, 2004) cannot appreciate the significance of the spatial layout of a map until he can move, jump, and shoot with adequate fluidity. Once the controls are learned and rudimentary movement and interaction no longer require conscious attention, space can be better appreciated and learned.

To give a simple example, players who have not engaged with the standard FPS control scheme of keyboard and mouse (*ASDW* keys to move forward, backward, and sideways, combined with the mouse to look around and aim) tend to struggle with navigating even the simplest of routes in the environment. New players learning to move in an empty *Counter-Strike*

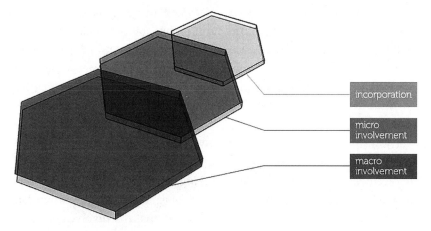

Figure 10.1
Incorporation and the player involvement model.

Figure 10.2
Navigating doors in *Counter-Strike: Source* (Valve Software, 2004).

map tend to have great difficulty going through open double doors, for instance. Since only one door is ajar and cannot be moved, players need to move forward and then diagonally either by pressing *W* and *S* simultaneously or pressing *W* and turning the mouse leftward as they go through. Instead, new players tend to go back and forth, left and right, while missing the gap in the door. While pressing the keys individually, the player often does not think of using the two-key combination, and the coordination of the left hand moving the avatar forward and the right hand turning her around proves too challenging at first. Until the double doors are passed through, they present an immediate barrier to action, which impedes the player's ability to translate movement into internalized space. The conscious attention directed toward learning the controls and navigating the space makes it difficult to enter a state of incorporation. The spatial and kinesthetic involvement dimensions thus combine frequently, with the internalization of the former being largely dependent on the internalization of the latter, creating the potential for incorporation.

Incorporation can be sustained but is often fleeting, slipping back into involvement the moment any dimension requires the player's full, conscious attention. Incorporation tends to become more intense when it is sustained for an extended period of time. Intrusions from sources unrelated to the game environment detract attention from the game, undermining involvement and thus incorporation.

This is not to say that incorporation requires sensory stimuli to arise solely from the game environment; it also accounts for input arising from outside the game environment, which the player can integrate into their game experience. For example, if you are playing *Left 4 Dead 2* (Valve Corporation, 2009) with friends in the same room and one of them cries out for help, this sound does not necessarily slip you out of a state of incorporation, since the call can be readily integrated into what is happening in the game environment. In fact, a call for help will most likely intensify the sense of incorporation because it fosters the absorption of the game environment into the player's consciousness more decisively by blending stimuli from the game world and from the immediate surroundings. Since incorporation is being conceived as an absorption into the immediate surroundings, rather than a dive into *another* space, it readily accounts for such instances of blending of stimuli from different environments. With incorporation, we do not need to view the game environment as a special, other space that requires protection from "real world" intrusion, as long as that intrusion can be integrated into the game experience.

Incorporation and the Four Challenges

In chapter 2, we discussed four important challenges faced by attempts to characterize the phenomenon we have been addressing here as *incorporation*. Specifically, we saw that any new concept must address the distinction between the senses of *involvement* and *transportation*; it must be specific to ergodic media such as digital games; it must avoid bias toward the power of technology to simply create such an experience; and it must address the multifaceted nature of the experience. The player involvement model and its extension to the concept of incorporation address all of these challenges.

First, our distinction between player *involvement* and its deeper realization in *incorporation* shows that our approach distinguishes between the sense of immersion as *involvement* and as transportation. The player involvement model describes the nature of a player's involvement with a game, presenting it as made up of multiple dimensions and as taking place in different phases, from the macro to the micro level. *Incorporation*, then, is our term for immersion as transportation, but, as we have seen, it expands upon this basic metaphor by including the view that the player is not merely transported to a virtual world, but also incorporates that world into her own consciousness in a dual process.

My definition and description of incorporation precludes its application to any non-ergodic media, such as movies or books. As we have seen, incorporation requires that a medium must specifically acknowledge the player's presence and agency within the virtual world. It is only in ergodic media that we find this kind of agency, and only in virtual environments of the sort we have discussed that such a presence in the virtual world is possible. A book or a movie is unable to acknowledge a reader's or viewer's presence, nor can it offer them agency, and so it cannot afford incorporation.

Note, however, that the acknowledgment of a player's presence is not sufficient for incorporation on its own—the player's subjective disposition must also be taken into account. The player involvement model and incorporation therefore account for the challenge of technological determinism. It is not merely a technical challenge to create incorporation; the player's role in shaping the experience is essential.

Finally, the problem of viewing the immersive experience as a monolithic phenomenon is addressed by the multidimensional nature of the player involvement model upon which incorporation is built. Rather than conceptualizing incorporation as a single form of experience in itself, the model describes it as arising from the internalized blending of six broad facets of the game experience. In this way, involvement and incorporation can be discussed not just in the abstract terms that have often been used up to this point, but according to very specific dimensions of experience that make them up.

Incorporation in Practice

When recalling memorable experiences in games, the majority of my research participants described holistic experiences in which consciousness and environment were integrated fluidly and accompanied by powerful emotions. In order to fully understand the concept of incorporation, we cannot only examine it as a term, but must see it as an actual experience. The following two examples of incorporation as described by two participants will serve to demonstrate this.

The Ruined Cathedral

There was a time when I was playing through *Guild Wars* [ArenaNet, 2005] ... it was in the war-torn parts of Ascalon. I was working through some ruins and I turned this corner, and came across this massive, ruined cathedral with this gorgeous stained-glass window that was mostly intact. I just stopped and stared at it. I worked my way around it as much as I could to see it from all angles and ended up on a rise a little above it, just watching it. I don't remember the time of day, but it might have been [around] sunset and I swore I could practically feel the breeze on my face and hear the wildlife. If I could pay to experience that in real life, I would. And I would pay *a lot*. It was a real moment for me, a real experience that I carry with me. (Rheric, *World of Warcraft*)

Rheric's account brings to the fore the intensity of emotion felt in such holistic incorporating experiences. If we were to remove the fantasy names from the anecdote, it would not be obvious to the reader that he was describing an experience in a virtual world. Rheric relates the event with strong connotations of inhabiting a place, emphasized by the synesthesiac addition of stimuli that were not part of the environment ("I could practically feel the breeze on my face and hear the wildlife"). Rheric's concluding sentence emphasizes the experiential significance of this event and the lack of separation between it and a nonmediated equivalent.

Figure 10.3
Guild Wars cathedral. Image courtesy of ArenaNet. © 2009 ArenaNet, Inc. All rights reserved. ArenaNet, Arena.Net and the ArenaNet logo are trademarks or registered trademarks of NCsoft Corporation in the U.S. and/or other countries.

Rheric's incorporating experience is clearly grounded in the internalization of spatial and kinesthetic involvement dimensions. The manner in which he describes the discovery of the cathedral revolves, at its heart, around the cyclical relationship between the two dimensions. As he navigates the game environment, Rheric repeatedly transforms his spatial ability into spatial knowledge. The work that must be performed in order to discover new spaces—the investment of effort in the kinesthetic dimension to continue to experience the space—is part of what makes the discovery of a new and unexpected space such as the cathedral so striking and affective.

The ludic dimension of the encounter then enhances this experience. We can infer from Rheric's description of the environment as "war-torn" and from our knowledge of the structure of *Guild Wars* itself that the area Rheric was exploring must have included resistance to his presence in the form of various enemies. Rheric's ludic and kinesthetic involvement combined in the formulation and execution of strategies and tactics for either eluding or confronting such opponents. Before he discovered the ruined cathedral, Rheric's movement was not just a matter of crossing an empty landscape, but of finding his way through it while meeting the surrounding dangers. The risk of attack and the challenges of navigation make a player more watchful and attentive to the environment. This, in turn, makes the sudden encounter with the ruined cathedral again more meaningful and affective. The contrast between his state of emotional readiness for combat and his sighting of the beautiful piece of architecture no doubt greatly enhanced Rheric's experience.

Finally, we must acknowledge the role of the narrative dimension in this powerfully affective and incorporating experience. The scripted narrative of *Guild Wars* creates a particular setting and background understanding of the inhabited world, both in terms of giving the environment a larger meaning and in creating a more exciting contextualization of the player's immediate actions. Rheric's alterbiography, simultaneously, is given a formidable plot twist when he comes across the cathedral, as it is a very dramatic moment. The affect experienced enhances the richness of alterbiography, once again, in combination with the other dimensions of involvement described above.

Narrative and spatial involvement thus combine in alterbiographical aspects related to travel and exploration. The engagement provided by open exploration is deepened by the added meaning provided by its embedding in the alterbiography. As Rheric's account indicates, such exploratory stories constitute some of the most powerfully engaging moments in MMOGs; indeed, a number of research participants related such stories with great excitement. An episode like Rheric's is powerfully gripping because it entails the blending of the five dimensions outlined here in a seamless fashion that is characteristic of incorporation.

Riding with the Ghosts of the Revolution

GotR [Ghosts of the Revolution] on the Emerald server host regular raids and require Team Speak. You should, just once, experience one of their raids! Just that one outfit can totally change the shape of a battle. Imagine ten-plus Magriders with gunners, all in Team Speak and moving as a cohesive unit. That would be a typical armor raid night. When they do an outfit raid, you have those Mags, air cover, grunts, and support all working together. There are designated people that talk between the different channels and coordinate the movement of as many as eight full platoons. One night they had over 85 people involved in a raid! It's a riot. And the sense of accomplishment is beyond anything I've ever seen in this game. To be able to say, GotR made this continent capture happen! It's a good feeling to know that you worked in tandem with that many other people and made something happen that might not have otherwise occurred. (Kumacho, *Planetside*)

Kumacho describes an altogether different form of incorporation. The incorporating experience in this case is facilitated by the collaboration among a large group of geographically distant people working together (shared and kinesthetic involvement) to overcome the opposition and achieve a common goal (ludic involvement). In order to carry out the collaborative plan, players participating need to communicate efficiently (shared involvement) to their leaders and subordinates while driving, shooting, and maneuvering (kinesthetic involvement) around the game world.

The performance of in-game tasks described by kinesthetic involvement becomes even more important when there are a number of people depending on your abilities to carry out the plan. If the driver of one of the Magrider tanks keeps bumping into trees and falling behind the advancing friendly column, the other players on board will get frustrated and the raid will be weakened. Because larger vehicles like the Galaxy air transport in *Planetside* carry more troops, even more avatar lives are at stake and thus more weight is placed upon the individual's skills.

Figure 10.4
Members of the Ghosts of the Revolution outfit during an aerial raid in *Planetside*
(Sony Online Entertainment, 2003). Image courtesy of Ghosts of the Revolution.

As the definition of incorporation states, the presence of other agents,
human or AI, within the environment is an important component of
habitation. When shared involvement is internalized, one's location within
the environment becomes further validated by the cohabitation. For
Kumacho, habitation of the environment is strongly influenced by the
number of people working together coherently. The group-defined goal of
capturing the opponent's base becomes a binding facilitator of inhabiting
the environment. The space inhabited becomes incorporated in this case,
because others are there also. In some senses, incorporation becomes viral.
When Kumacho hears a request to turn the tank he and his teammates are
in due west to face an oncoming threat, it contributes to his sense of
inhabiting the place because the habitation becomes shared.

The ability to work with others and contribute to the collective to which
one belongs, in this case the GotR outfit of which Kumacho was part, is
dependent on knowledge of one's ludic and kinesthetic involvement.
Based on their knowledge of the situation around them, players start

inferring ways in which to aid and work with others without needing to communicate in great detail. This ability to read the situation around them is dependent on knowledge of the game's rule systems. Thus, a degree of internalization of the ludic involvement dimension is displayed, including an understanding of the hierarchy of goals as the collective players and the game system shape them.

The complex operation in which Kumacho participated presents GotR with a series of goals organized around the central overarching goal of eliminating the target of the raid. A lot of things can go wrong in the pursuit of such a goal. Aside from the logistical challenges of coordinating a large number of players, there are also the challenges imposed by the opposing factions. Because each individual player's skill and performance impact the outcome of the raid, both the troops on the ground and their leaders must alter plans on the fly. But the more demanding any instance is on the kinesthethic and shared dimensions, the fewer attentional resources are available for planning (ludic involvement).

As Kumacho's account describes, the presence of others also opens up the potential of greater emotional affect. For Kumacho, the momentum provided by successful coordination with the rest of his outfit, which we can describe as internalized shared involvement, results in a deepening of emotional impact, or affective involvement. For Kumacho and the other members of the GotR outfit's raid, affective involvement becomes intensified because the situation is both strongly collaborative and competitive, with as many as a few hundred players on each side vying for victory. If, after four hours of besieging a fortress without progress, a team of coordinated players, like Kumacho's GotR outfit, managed to infiltrate the fortress and turn the tide of battle, the excitement of their exploits would be greatly enhanced by the fact that they are influencing so many other players, both on their own side and on the opposing faction.

Though Rheric's and Kumacho's accounts are quite different in nature, they both express the evocative power that instances of incorporation can create. Incorporation offers a rich understanding of the phenomenon of virtual environment habitation that, unlike the metaphors of presence and immersion, is multifaceted, media-specific, and precise in its formulation. It also has the advantage of being built on a robust model that describes the various dimensions of involvement that lead up to it along a temporal axis, ranging from off-line involvement to conscious attention during gameplay and culminating in the blending of involvement dimensions that result in incorporation. Finally, with the concept of incorporation, we no longer need to draw a strict line of demarcation between stimuli emerging from the virtual environment and stimuli emerging from the physical world, for the emphasis is placed on the internally constructed consciousness of the individual. Thus, incorporation allows us to move beyond the notion of virtual environments as experientially separate otherworlds and to treat them instead as domains continuous with the media-saturated reality of everyday life.

Conclusion

During the first few hours of playing a new digital game, you adapt yourself to the game's mechanics, physics, and rules. You take the first uncertain steps in the unexplored spaces, experiment with running, jumping, leaning around corners. The boundaries of character creation are prodded and initial strategies of virtual world domination are formed. In geographically rich game worlds, you're likely to explore the first areas slowly and thoroughly, until you form a cognitive map of the layout of the world, country, region, or city. The background story is delivered, told through scripted narrative channels. In multiplayer games, the first lines of communication are typed in chat boxes; in a multiplayer FPS setting, chances are you will be hearing cries of "Follow me," "Cover the corridor," and "Come on, RUSHHH, you noob" from seasoned players. It is all very confusing at first, but slowly you learn and internalize the ways of the (virtual) world. Later, when the controls are second nature to you, and the environment of the game has become a familiar space for you to move within, you find yourself feeling as though you exist within the world. You become accustomed to your digital manifestation of self.

This phenomenon might be described as an instance of "the willing suspension of disbelief," and often is. However, such a description may not apply well to games. The comparison may be invidious, but a suspension of disbelief is probably needed more when our engagement is limited to interpreting what we are given by the writer of a text or a film's script than in digital games, where belief (if the term still applies at all) is created through action, movement, navigation, communication, and other forms of interaction. How, indeed, does one *suspend* disbelief when one is so *extended*, physically as well as affectively and imaginatively, into a game?

Whereas the notion of suspension of disbelief is primarily dependent on the imaginative faculty being stimulated by words or images, belief in game worlds is created and sustained through the cybernetic looping of simulation and imagination. The landscapes, characters, objects, and events that make up game worlds become real in players' minds not only because they appeal to the imagination (although this is a crucial requisite), but also because that imagination has an existence outside the player's mind in the simulated and digitally materialized world generated by the computer. An important part of that simulation is the ability to act within the game environment and thus to feel that one has a tangible presence therein. As we learn to act within a game world, our actions become more fluid and complex; we come to know the environment's affordances and geography. When we combine the imagined habitation of the game environment with the validation of our location within it, we are not merely suspending disbelief but incorporating the computer-generated world and its inhabitants into our consciousness. The game environment shifts from being a space where we manipulate (external) digital puppets to a place continuous with the haunts of our everyday lives. For most players, the merging of virtual spaces into actual ones is a normal part of gaming. It is neither an alarming sign of addiction nor an indication of irresponsible escapism (or at least, not necessarily so), but an experiential state characteristic of engagement with certain types of digital games; and one with a phenomenology to it that has specificities that bear further scrutiny.

Implications

This book has sought to reposition the interpretation of experiences afforded by digital games and their social and cultural significance by reframing the relationship between the media object, the player, and the social and cultural contexts that envelop both and are differently inscribed in each. This tripartite relationship challenges typical binaries such as game/nongame, work/play, and real/virtual by emphasizing the mediated interactivity of player and virtual environments. Players are productively bound to these virtual environments via social and cultural codes that are designed into the game; these codes are also acquired through the practice of remediated everyday life by the player.

The problematic binaries mentioned above also underpin the numerous conceptualizations of immersion and presence. Both of these metaphors are taken to imply a physical reality that is replaced with a virtual world in a here/there dichotomy that misrepresents our mode of being in the everyday world. The metaphor of incorporation allows us to avoid such a dichotomous relationship: it expresses the phenomenon of immersion or presence as assimilation into consciousness of the game world in a manner that is coextensive with our being in the physical world. This does not imply that the two are equal, but rather that the physical and the virtual are both aspects of what we perceive as real. It is crucial that we distinguish between considerations of the actual nature of reality and human perception thereof. When our concern is the human perception of reality, as is the case in this book, we should not take the physical real as the stable point of comparison to which virtual phenomena are measured up. The virtual is, rather, a crucial aspect of contemporary reality.

As was anticipated at the start of the book, the definition of games as a discrete, holistic category of media objects is challenging to sustain in light of the increasing variety of objects and activities that fall under the label, at least in popular discourse. This is not necessarily a problem, as long as researchers are aware of the variety of members in the game family and take care to clarify which subset of this broad family their research is focused on. Needless to say, some delimitation *is* in order, if the analysis in question is to move beyond vague generalizations. This has become particularly problematic with the advent of games in virtual environments, since virtual environments allow for activities in their spaces that cannot be easily folded into the field of games in general. Consequently, as the model has shown, we should not assume that every engagement with the objects we call games results in a specific form of mental attitude, most often denoted by the concept of play. Accurate analysis of player involvement will remain limited if we do not decouple the notion of play from the media objects we currently call games.

Of course, any model for the analysis of player experience will have some inadequacies. If we consider Gérard Genette's excellent *Narrative Discourse* (1980), we are immediately reminded that even the most incisive of models—in this case, the analysis of narrative time in literary texts, based on Marcel Proust's incredibly complex *À la recherche du temps perdu*— required a revision and a response to criticism, which was later published as *Narrative Discourse Revisited* (Genette, 1988). This is not to suggest that this volume will immediately necessitate a revision, but some acknowledgment of the particular difficulties that games and virtual environments present to theoretical modeling may well be in order here. Models are analytical constructs that by their very nature may lack the suppleness and expansiveness to adequately account for complex phenomena, particularly if they touch upon issues of human subjectivity and consciousness, as is the case with this work. There will always be aspects of the phenomenon in question that will remain challenging to classify and label. This concern reminds us of the tribulations undergone by structuralism, which was challenged by poststructuralism precisely because of the latter's greater attentiveness to those texts and the elements within them that undo the work of taxonomy and of modeling itself.

Even with all this in mind, it is still worthwhile, and indeed necessary, to begin our mapping of a complex phenomenon, such as the one discussed in this book, through structural models. These models can then be elaborated upon, refined, and eventually revised until such a point when the scope of our research outgrows these analytical tools. Going back to literary theory as an example, we can safely say that the analytical sophistication of poststructuralism has been made possible by the existence of structuralism and formalism before it, as well as by the evolution of literary texts that necessitated such analysis. I would argue that games, and indeed their analysis, have yet to reach that level of textual sophistication. This book is, therefore, a first, considered step on my part—doubtless flawed, but nevertheless essential at this stage within the critique of games— toward that kind of more evolved and sensitive modeling. It is an attempt at a refinement of the concepts and models that have thus far described player involvement and immersion.

Imagining Worlds into Being

Digital games foster a close relationship between the collective social, cultural, aesthetic, and technical resources of the design team and the game-playing experience. This relationship is complicated in multiplayer games and virtual worlds because of the participation of other players who affect one another's subjective experience of the game environment. Human players and, to a more limited degree, AI behavior modulate what designers intended or expected players to experience, even if human players have an important role in describing the broad parameters of the experience: its setting, ambiance, characters, physics, game rules, tempo, and so on.

Fictional worlds have always been meticulously designed to allure us into inhabiting them. We have gone from sharing our imagined worlds through images and the spoken word delivered to a proximal community to distributing these worlds at greater distances through writing and later print. Mechanical technologies provided the possibility of creating moving images of these worlds. With the advent of networked digital technologies, we now have the ability to simulate these worlds and share them across the globe instantly. Today we are able to inhabit each other's creations and witness others inhabiting our own creations. As this book has shown, the variety of experiences that virtual worlds make available is only limited by the creativity of the designers and the resources they have at hand. The sensation of inhabiting a virtual world thus takes on a variety of forms suiting a range of tastes. From the calm pleasures of watching a beautiful sunset with your friends, to the adrenaline rush of a competitive FPS match or the exhilaration of piloting a speeding jet through a gorge, incorporation is an experience that is compelling for designers, consumers, and technologists alike. Incorporation is becoming an increasingly sought-after experience that is most readily provided by virtual game environments. The increasing amount and variety of these media objects is striking evidence of the complex human needs and interests which such simulations are able to engage with and satisfy.

Appendix: A Tale of Two Worlds

For the sake of readers who have not experienced the games firsthand, this appendix gives a brief overview of *World of Warcraft* (Blizzard Entertainment, 2004) and *Planetside* (Sony Online Entertainment, 2003), the two worlds inhabited by my research participants. These descriptions are meant to highlight specific features of the two worlds and to provide readers unfamiliar with massively multiplayer on-line games (MMOGs) with an overall description of the two genres: massively multiplayer on-line role-playing game (MMORPGs), the genre of *World of Warcraft*; and massively multiplayer on-line first-person shooter games (MMOFPS), the genre of *Planetside*. The following sections will discuss these genres as well as the spatial and geographical design of the two games.

Genres

Although both games fall under the broad category of massively multi-player on-line games, *World of Warcraft* is more specifically a massively multiplayer on-line role-playing game, while *Planetside* is a massively multiplayer on-line first-person shooter. The *role-playing* tag in MMORPGs can be somewhat misleading to those familiar with the pen-and-paper variants, such as *Dungeons and Dragons* (Arneson and Gygax, 1974), in which players perform as characters through verbal communication and gestures. Since these performances can range from speaking and acting in or out of character according to changes in tone of voice, accent, and gestures, playing these pen-and-paper RPGs can amount to giving a full-blown performance of the role enacted (Fine, 1983; Mackay, 2001). For a variety of reasons, the vast majority of MMORPG players do not tend to filter their actions through a particular persona, with the exception of players who inhabit incarnations of the virtual world that supports role playing, often called role-playing servers. Rather than requiring players to play fictional characters or personas, MMORPGs mostly differentiate the broader genre of worlds from more particular ones like those of MMOFPSs and MMORTS (massively multiplayer on-line real-time strategy) games, as well as drawing on the mechanics of their pen-and-paper forebears.

MMOFPSs can have very similar traits to MMORPGs, but their focus is on real-time action that most often requires a first-person perspective. Currently, the main feature that distinguishes MMOFPSs from other MMOGs is the combat system. As their name implies, MMOFPSs tend to be set in environments that facilitate long-range combat, such as *World War II Online* (Corner Rat Software and Playnet, 2001), *Face of Mankind* (Duplex Systems, 2006), *Planetside*, or the upcoming *Huxley* (Webzen, forthcoming). The game mechanics in MMOFPSs tend to require more *twitch*-based skills, with players employing more rapid hand-eye coordination and reflexes than in MMOGs and MMORTSs.

On the other hand, MMOFPSs also accommodate nonfighting playing styles in the form of *support classes* like medics, engineers, and transport drivers. They may also include character progression, classes, and missions or quests similar to what is seen in MMORPGs. MMORPGs, for their part, can often be played from a first-person point of view and can involve real-time combat. The dividing line between these two is not set in stone, but tends to be a product of the world's setting: the more guns available, the more likely the game will called an MMOFPS.

Topographies

World of Warcraft is set in Azeroth, a fantasy world made up of ninety-six regions spread over two continents. Each of the regions is controlled by one of the two factions, the Alliance or the Horde, or is a *contested area*, with settlements of each faction present. Regions have a unique aesthetic style and are painted in distinct palettes. The individuality this gives to each region enhances the sense of exploratory novelty as the area is encountered. The transitions from region to region tend to be unnaturally abrupt, with a clearly locatable few meters of land containing a dramatic blend of colors from the two regions' palettes. Players do not view this lack of mimetic realism negatively, however, and in fact it seems to work as an

Figure 12.1
Two regions in *World of Warcraft* (Blizzard Entertainment, 2004).

effective mechanism to differentiate the areas and make the world feel larger than it actually is.

My research participants did not seem to be bothered by the lack of geographical coherence of the world. Geological principles are thrown to the wind, with the lush Feathermoon region lying a few hundred meters from the barren desert of Desolace. A brief look at the Google map version of *World of Warcraft* (*World of Warcraft* Map, 2006) makes clear Azeroth's quiltlike terrain. Rather than going for subtle shifts in terrain generated by geological principles, as is the case in MMOGs like *Dark and Light* (NP Cube, 2006), Blizzard opted for a consistency of aesthetic implementation rather than scientific realism. Blizzard has also cleverly manipulated methods of creating impassable boundaries like mountains, forests, walls, and shores that channel players along designed pathways. This also makes the regions seem larger than they are, as the channels seem to stretch further than the traversable terrain. These techniques are common in digital game environments where, for example, a city seems to trail off into the distance, but the area that can be accessed consists of only a few streets, as in *Half Life 2* (Valve Software, 2004).

The world of *Planetside*, Auraxis, consists of ten continents and five caverns. Each continent has a particular terrain theme and weather cycle, but the landscape is visually generic, and aside from weather effects, the ambient sounds are negligible. Participants commented on their difficulties with differentiating the continents, even after having played regularly for over a year. Even though the terrain and climate is different for each continent, the visual difference between them is not as distinct as that of the regions in Azeroth. Auraxis is an aesthetically blander place than Azeroth, with few distinguishing features to explore. The lack of AI-based agents also makes areas without human players incredibly barren. Aside from towers and facilities, there are no urban areas in *Planetside*, an element that was raised by the majority of players as something they would like to see in the world.

Planetside has no safe areas aside from the starting factional continent called Sanctuary. Unlike that in *World of Warcraft*, the *Planetside* world does not have any calm, social spaces like marketplaces, auction houses, taverns, and the like. Thus, socializing is restricted to collaboration with comrades from one's outfit; it is rare to see gatherings of friends socializing as one so often does in the urban centers, and especially the auction houses, of *World of Warcraft*. The lack of social spaces in *Planetside* can also be attributed to the fact that items are obtained from equipment terminals, leaving no room for a persistent, player-driven economy. Conversely, in *World of Warcraft*, the demand and supply of items creates a marketplace feel to some areas, with players advertising their wares on the trade channel and needing to be close to buyers or sellers in order to secure a transaction. This geographic proximity is important, as one-to-one transactions are often preferable to those at the auction house, particularly when advantageous deals need to be snatched or a particular item or material is required immediately. Finally, *Planetside* also lacks socializing spaces due to its fast-paced action and pervasive PvP (player-versus-player) action. Players need to be on their guard at all times, as one cannot know when a shock assault or infiltration mission is being launched by one of the opposing factions. Naturally, *Planetside* attracts more action-minded players, with socializing often being left to outfit chat channels or directed more specifically to collaboration in immediate tactical operations.

The Lore

The backstory of *World of Warcraft* builds on the highly popular *Warcraft* trilogy of RTS games. This popular precedent eased entry into its setting for a number of players who had played the earlier games and were familiar with the lore portrayed in those titles. Nevertheless, the depth of geography, history, and mythos are nowhere near as rich as literary and tabletop

Figure 12.2
Two regions in *Planetside* (Sony Online Entertainment, 2003).

RPG worlds such as Tolkien's Middle-earth or Greenwood's Forgotten Realms. Blizzard's world seems to have just enough lore to keep those interested in it occupied, without displaying a sense of the depth and long-term evolution characteristic of secondary worlds in fantasy literature or some pen-and-paper role-playing game settings. This has a lot to do with the lore's lineage, for the initial *Warcraft* game pitted orcs against humans in a generic environment, not a setting conducive to a deep secondary world:

I'm curious, but I won't go out of my way for a lot of background. Orcs versus Humans. ;) *World of Warcraft* tries to have some sort of story background, but it's just so few and far between that you encounter it. Some of the quests have interesting mini-stories, but the overall game theme is a bit weak. Well, with new quests, I read the quest descriptions and tried to understand what was needed. Now, I just click the Accept button and look for the bare minimum needed to be accomplished to get the reward. (Haelvon, *Planetside*)

Haelvon's attitude toward the lore was common among the research participants; this disinterest is reinforced by the fact that the games do not require players to engage at all with the lore of the world in order to progress or to be involved within its societies. Most participants did not find it rich enough to capture their attention. This ambivalence to the designed narrative of the game was compounded by the fact that the story delivered by quest texts can be entirely ignored without hindering progress in the quest: "I don't even read the quests … just look at what it takes to mark them as complete. It's not real and doesn't really impress or interest me.

I don't think it really makes any difference if you know or follow the storyline" (Nombril, *World of Warcraft*).

Unlike *World of Warcraft*, which is set in a medieval, swords-and-sorcery-style fantasy world, *Planetside* is set in a futuristic era of space colonization gone wrong. A contingent of humans is stuck on the planet Auraxis. Their disconnection from the rest of the Terran Empire has splintered the army into three factions: the Terran Empire, the New Conglomerate, and the Vanu Sovereignty. Players must pledge allegiance to one of these three factions and lend arms to the never-ending battle against the other two.

Clone Worlds

World of Warcraft's Azeroth exists on thousands of servers around the world. Each new character must be created on a specific server. Each server runs a clone of Azeroth that develops its own specific social and economic history through time. These thousands of Azeroth clones are divided by geographic region: America, Australasia, Europe, and Asia. Once the *World of Warcraft* application is bought within one of these regions, the player is restricted to playing on servers of that region. Players who want to play with friends in distant geographical locations may thus find it impossible to get together inside *World of Warcraft*. As some participants have noted, the reason they stayed in a particular virtual world or moved to a new one was to keep previous connections with other players alive when they move to another continent:

I have changed games several times to stay with people that I have gamed with previously. Even when we moved to England, I continued to play with people that I had gamed with before. Although we didn't have to "leave" those friends behind, I think that I would have felt bereft and more "alone" in a new country and in unfamiliar surroundings if I had had to do so! A group of friends that I have played with online together for approximately four years has pretty much stayed together, and we live across the U.S. and Canada. Currently we have played three different online games together (SWG, *World of Warcraft*, and now DDO). Although we keep in touch primarily in game, we have all at some point or another communicated via email, instant messenger, and voice chat, even though we have never met any of these people in person. (Ananke, *World of Warcraft*)

Planetside's Auraxis has only three clone servers running: one on the U.S. West Coast, the other on the East Coast, and one in Europe. There are also no regional restrictions with regard to which servers players can log on to.

Customization

In *World of Warcraft*, players choose a race and character class, give their character a name, and then are able to alter the look of their character. Yet *World of Warcraft*'s character customization is more restricted than that of other MMORPGs. Aside from a few options, such as changing a character's hairstyle, skin color, and face shape, players have very little opportunity to make their avatar look unique. All the participants interviewed commented on this factor, mostly in negative terms. The ability to create a unique avatar is clearly an important factor for players, and *World of Warcraft* is underwhelming in this department: "Character customization is limited. There are not nearly enough options that enable you to differentiate your character from others. No real way to make your character an individual. ... Aside from the name and their gear, what does one level 60 rogue have that's different from another at the end of the day? Dagger build, Sword build, sheesh!" (Inauro, *World of Warcraft*).

As Inauro notes, one way of differentiating one's character from others in *World of Warcraft* is through the acquisition of rare clothing, armor, and other items. In many ways, *World of Warcraft* literalizes the idea that you are what you own. Differentiation from others is dependent not on one's creativity but on the investment of long hours in the world accumulating riches and goods, or the expenditure of real money on *World of Warcraft* gold and other items. In *World of Warcraft* you literally are what you wear.

Planetside suffers from a greater lack of avatar customization than *World of Warcraft*. The only changes to the preset avatar that can be made are to the character's facial features, hair, and general head shape. Otherwise, all players look similar to others from their faction. As the character progresses, though, the possible combination of skills makes for a greater variety of character compositions and play styles. Unlike most MMOGs, *Planetside* has no character classes. Each character is defined by the chosen skills and by how the player combines those skills during play. Some skill groups enhance the chances of doing particular actions, while others enable completely new abilities that would not otherwise be available.

Mechanics

Although *Planetside* allows players to connect to any of its three servers, geographical distance from the game servers is more of a problem than it is in *World of Warcraft*. Geographical distance and quality of network connection determine the amount of latency between the player and the server. In the case of *World of Warcraft*, a 300-millisecond latency will not create a great functional difference, as character actions consist of sequences of moves activated by key strokes that are queued by the server and then implemented in the game sequentially. In *Planetside*, however, actions are performed in real time. If there is a slight delay in server response time, you might be shooting at things that were there half a second ago, causing you to miss. In *World of Warcraft*, once a character has been ordered to attack an enemy, the character will automatically follow the targeted enemy, so if the enemy is not where they appear to be due to latency, your character is not there either, and has followed the mob and executed the strike anyway. In fact, when clogs in on-line traffic occur, the screen might freeze, but your character will still perform the queued sequence of attacks. In this way, a *World of Warcraft* character can retain a certain number of orders and perform them regardless of continued player attention or input. In *Planetside*, the character is more closely aligned with the player's input and thus does not perform any automatic actions. To highlight this important difference, let us envision a similar situation in the two worlds.

A player is wandering through a dense forest at night in the game. The darkness makes it harder for the player (and supposedly the character) to see what is lurking in the distance. Both *World of Warcraft* and *Planetside* players will be cautious and look for enemies hiding in the vicinity. The *Planetside* character's and player's perception of environment are more closely aligned. If the player cannot see the enemy, neither can the character. If the player does not notice a figure running from tree to tree to her left, the character did not see it either. In *World of Warcraft*, the player can hit the Tab button repeatedly to target enemies in the vicinity. The character may target a creature the player cannot see, and even attack that target if it hasn't moved out of range. All the player needs to do is hit the right command button and the character executes the action. In contrast, in *Planetside*, the player needs to see the target, align his crosshairs, and click on the mouse to shoot.

All of the *World of Warcraft* participants stated that they play in third-person perspective so that they are able to have a wider field of vision than their character, including areas like the rear, which the character cannot see. Among *Planetside* participants, there was unanimous agreement that it is only possible to play from a first-person perspective, even though a third-person one is available. This is simply because players need to be able to aim or operate objects themselves rather than have the character aim or perform actions for them.

As a result, every small change in the environmental situation has a far stronger impact on a *Planetside* player than on a *World of Warcraft* player. The heavier the rain, the more occluded the line of sight will be, making it easier for characters wearing infiltrator suits to remain undetected, and harder for characters shooting at a long distance to mark their target. The presence or lack of light raises similar issues. Use of cover and elevation are crucial in *Planetside*, while in *World of Warcraft* the only terrain considerations are distance from computer-controlled or player targets and passable/impassable areas.

Like the majority of MMORPGs, *World of Warcraft* operates on a level-based structure. Every entity, item, and location is assigned a level value. A character attains higher levels by completing quests, killing monsters, and participating in specially designated player-versus-player (PvP) areas called battlegrounds, which exist as instanced minigames within the world rather than being a persistent part of it.[1] The higher the difference in level between the character and the entity defeated, the more experience points (XP) are gained. Characters are also awarded a token reward of XP when they discover new areas, although the amount is not enough to make exploration a specific incentive to develop one's character. Items and skills require a minimum level to be used, as do more efficient means of travel.

Each area in the world is inhabited by entities of a certain level. These entities (normally called *mobs*, short for *mobiles*) attack players when they come within a radius whose size depends on the difference in levels between the character and the entity. The more the difference in levels favors the entity,[2] the larger the radius at which it will attack players. In *World of Warcraft*, no matter how skilled one is, trying to defeat a mob or player eight to ten or more levels higher is almost impossible. There is a close relationship between a character's level, her freedom to travel in the world, and her power over lower-level characters. Thus, players need to invest time in their characters if they wish to be able to experience the greater part of the world safely.

Quests in *World of Warcraft* tend to be fairly straightforward affairs. The player is usually asked to deliver something to someone, kill X number of creatures, or kill X number of creatures until she finds Y items on their dead bodies. There are a few more interesting quests peppered throughout the game, but most are structured simply. As participants have commented, it is often unnecessary to read the actual text of the quest. Instead, all the players need are a target, an amount, and a location. As the majority of participants have confirmed, the narratives presented in the quests and the overall background of the world are often ignored while players deal with quests functionally. This functional perspective on questing is the direct result of a system that requires characters to acquire XP and ever higher levels in order to progress, thus rewarding efficiency over creativity. In order to boost efficiency, most players speed up the quest process by using external Web sites where other players have posted information about the relevant quests. The most common type of help given comes in the form of map coordinates showing where the quest objects or mobs are located. Using help from external sources is viewed by most players as being an accepted part of the game itself. Players rationalize that they use such sites because most quests are repetitive in structure, and there is no satisfaction in solving them. Although most MMORPGs have level structures of some sort, how much these levels affect the overall possibilities for action in the world can vary. *World of Warcraft* places great emphasis on defining characters by their levels. In PvP servers, characters who are of a much lower level than their attacker have no chance of competing against their assailant, which can lead to griefing. *Griefing* refers to a situation in which one character kills another character who is below her level and thus would not have a chance of surviving the attack and often, as is the case in *World of Warcraft*, without receiving any form of reward for the action. Some griefers then *corpse camp* their victims, staying close to the dead body of the victim in such a way as to slay them when they return to claim it. Without intervention from the victim's faction, the griefing character can stop the victim from resurrecting by slaying them as soon as they come back to life, locking them out of the game indefinitely. Although there are no strict rules against such practices, and some players claim that they are within their rights because they are part of the PvP mechanics, griefing and corpse camping are socially frowned upon.

Planetside's character development system is also based on levels, but, unlike *World of Warcraft*, attaining a new level does not change the character's attributes, skills, or powers. It only means that the character earns more points to assign to proficiency in a particular weapon category or profession, such as hacker, medic, or engineer. The higher a character's level, the more skill, vehicle, and weapon categories (called certifications) are available to them. A starting character might be able to use three types of weapons and ride a bike, while an advanced character may be able to use eight types of weapons, control tanks and aircraft, and use medical devices on themselves and others.

Nevertheless, the starting character can still compete or collaborate with an advanced character, providing she has the right timing, tactics, and skill. Advancement in *Planetside* results in a more flexible makeup of abilities, while in *World of Warcraft* a character's level determines which groups she can join and how much absolute power she wields. In *Planetside*, no matter how strong and skillful a character is, a handful of beginning characters will generally defeat him through superior numbers. It is a perilous world for all: disputes between teammates can erupt in duels, and the calmest seeming tract of land can have enemies waiting in ambush.

Unlike in *World of Warcraft*, *Planetside* players do not choose a particular race or class for their avatar. Instead, their capabilities are determined by the combination of certifications chosen and their progression based on command experience, which is accumulated by leading squads and platoons. Although *Planetside* may seem more combat-oriented, *World of Warcraft* characters are also all essentially fighters with varying capabilities. In *Planetside*, characters can be pure noncombatants and can focus on serving as engineers, medics, hackers, or drivers. Engineers repair vehicles, aircraft, turrets, and other facility equipment. Medics restore lost health to wounded infantry and can bring fallen comrades back to life. Hackers are key to capturing facilities and towers and can also hack enemy vehicles and terminals, turning the tide of battle through tactical manipulation. Drivers can choose from a combination of land, sea, and air vehicles; most often they focus on mass transport vehicles that enable airborne attacks and other tactical strategies. Players can choose any combination of skills, but those dedicated to support roles can normally use only a limited range of weaponry, as their certification points require focusing on their primary skills.

Unlike *World of Warcraft*, *Planetside* rewards support roles by giving support experience points as well as a share of battle experience points (the more general XP in *Planetside*) out of any experience points that are accumulated by healed, repaired, or driven units. This creates an XP system that rewards collaboration even with comrades outside one's squad and platoon. The possible combinations increase the scope and need for teamwork, as player-created tactical tasks usually require a certain number of specialized roles, without which the group will be ineffective. For example, in *World of Warcraft*, although there are characters called rogues who specialize in detecting traps and opening locks, there are no quests that require their specific expertise, whereas in *Planetside* a sabotage mission aimed at hacking an enemy facility requires the skills of a hacker. Other characters cannot replace this role, no matter how strong their firepower and fighting skill is. *Planetside* thus gives a more functional role to specialized characters, giving players a stronger sense of agency and of belonging to their faction and platoon:

I like to be useful to others, so I choose certifications that allow me to fulfill a tactically helpful role. First and foremost, I take an active part in leading platoons. The tactical side of *Planetside* is something that is quite unique. To fulfill this role, I make sure I can fly transport aircraft to redeploy troops, engineering to repair others and set up defenses and infiltration suit/wraith combo so I can move around rapidly and undetected. Like that I can give feedback to my troops and coordinate attacks with other leaders. (Baal, *Planetside*)

Another major difference between *Planetside* and *World of Warcraft* is the scope and variability of situations. *Planetside* is a more open world without any structured missions or AI entities. The events of the world are shaped completely by the players, making for varying and unpredictable situations that cannot be anticipated by individual players or even high-level generals. This has its advantages and disadvantages. The very nature of player-generated situations often creates unbalanced scenarios. It is not uncommon to be defending a facility against an army that outnumbers the defenders by as much as five to one. But that is also part of the pleasure of these situations: The aim is not necessarily to win the battle, but to hold off the enemy, or whatever spin the participants in that situation decide to give it. Often it is inevitable that the facility will fall, but the desperate defense is something players enjoy; otherwise they would simply abandon the impossible task. *Planetside* puts considerable emphasis on mass collaboration, and participants stressed the attraction of a world that allows for mass battles between globally distributed players.

Planetside

Where *World of Warcraft* minimizes unpredictable situations by virtue of its tightly scripted and restrictive design, *Planetside* provides the game mechanics, rules, and uninhabited areas for players to create their own events, allowing a more open-ended experience. At times, this lack of regulation in *Planetside* can lead to frustrating situations that may put players off, as one is dependent solely on others being present to achieve anything in the world. Conversely, *World of Warcraft* gives far more scope for solo activities, ranging from quests to crafting to playing the auction house for economic gain. That said, *World of Warcraft* does not give much scope for players to feel like they can make a difference in the world of the game. Mobs respawn in the same locations, nonplayer characters (NPCs) give the same quests day after day,3 and the resources gathered or mined in a particular location pop back into existence a few minutes later. *Planetside* gives a greater sense of agency because players can change the course of a battle by their efforts, particularly if these efforts are concerted tactical ones and are well executed:

I was in D2A [the Death to All outfit in *Planetside*] when there were five servers: two each for East and West coast. They were very organized and now I feel a certain devotion to the outfit.

It used to be a requirement to be in teamspeak [a voice-over IP program that allows groups of players to talk to each other simultaneously]. If you would like to see a really well-organized outfit, make a VS [Vanu Sovereignty, one of the three factions in *Planetside*] on Emerald and join GoTR [the Ghosts of the Revolution outfit]. They hold regular raids and require teamspeak. You should, just once, experience one of their raids! Just that one outfit can totally change the shape of a battle. (Kumacho, *Planetside*)

It is hard to think of equivalent situations in *World of Warcraft* that highlight the exhilaration and sense of making a difference in the world like that Kumacho describes. Even if, on a PvP server, one faction captures an outpost or village belonging to the opposition, NPCs will keep respawning and it will be hard to hold the place for any length of time. Even if it is held, there is no in-world reward or advantage: when the aggressors leave the area, it will resume its previous existence, as if nothing had happened. In *Planetside*, taking a whole continent makes it very difficult for other factions to regain a foothold there due to the lack of spawn points and accumulated benefits that connected facilities give to their controlling faction.

Although these two worlds have been around for some time, at the time of writing this book each represents the prototype of its own genre: *World of Warcraft* for MMORPGs and *Planetside* for MMOFPSs. Between them they encompass a wide variety of gameplay forms that are present in a variety of other genres while also adding the *massively multiplayer* element. For this reason they have proven to be great sites for researching and developing a new model for player involvement. Future application of the model to other genres of games will update and expand upon the solid foundational research these two worlds have provided.

Notes

Introduction

1. Throughout the book, I use the term *digital game* rather than *computer game* or *video game* in order to preserve the generality of application rather than privileging any particular hardware platform.

2. Although the conceptual model forwarded in this book is also applicable to the related family of graphic virtual environments, I have retained the use of the term *player* instead of *user* both to emphasize the gaming roots of this work and to highlight the internal, subjective dimension that lies at the heart of any textual engagement.

Chapter 2

1. The original technical term proposed by Minsky in 1980 was "telepresence," but in contemporary discussions this has been shortened to "presence."

Chapter 4

1. Evita is one of my research participants. Direct quotations from my interviews with research participants will be cited in the text in this format (screen name, game name).

2. Rewards for goal achievement will be discussed in greater detail in chapter 9.

3. A particular aspect of the game (such as a character class, spell, weapon, vehicle, or skill) becomes "nerfed" when the game designers alter the characteristics of that object to make it less powerful, normally because it is seen as being too powerful or balance-tipping.

Chapter 5

1. Scripted narrative will be discussed further in chapter 7.

2. By "technical boundaries," I am referring to both the coded environmental properties and the game rules.

3. See the appendix for a description of the two MMOGs.

Chapter 6

1. Most MMOGs have a communication channel that reaches the local region or entire world (depending on design) separate from the close proximity "say" or "yell" channel.

2. *EVE Online*'s equivalent of a guild.

3. Detailed coverage of the event can be found at http://eve.klaki.net/heist

4. "Instances" are areas created specifically for the group and are run as pocket areas in the world which no other users outside the group can enter. These often require the coordination of anywhere between five to forty players to complete as well as a lot of preparation prior to the "run."

5. MMOGs designed around a level structure (see chapter 4 for a more detailed discussion of this) are frequently considered to have two main phases of the game: the progression from level 1 to the maximum level (60 in the case of *World of Warcraft*) and the qualitatively different activities available to characters of maximum level, often called the *end-game*.

6. A *clan* is a group of players that play together in tournaments which can be organized on line or at local LANs.

7. Some MMOGs exist on different servers. These may also operate on different game or social mechanics. For example, PvP (player versus player) servers in *World of Warcraft* allow characters of one faction to attack any character from other factions at any time. On PvE (player versus environment) servers, characters may be attacked only if they have their PvP flag on, which can occur if they engage in combat with other "flagged" characters or if they attack an NPC (non-playing character) of the opposing faction. There are also role-playing servers, where there is more emphasis on speaking in character (reminiscent of MMOG's pen-and-paper cousins).

Chapter 7

1. Cut scenes are animated sequences which the player has little or no control over. They are used to deliver sections of scripted narrative to the player and thus further

the story. They often also function as a form of reward for completing a task or level in the game.

2. A quick-time event is similar to a cut scene in that it delivers a segment of scripted narrative to the player, but it involves an element of gameplay. Usually this comes in the form of pressing the correct buttons or joypad movements when these appear on screen.

3. Tabletop RPGs are useful in such analyses because the mechanical workings of the system are transparent to the players and game masters, since they are expressed in numbers that are made meaningful through the cognitive interpretation of the rule system. Digital games hide these calculations within their layers of code, making them accessible only to those who can dissect and interpret the code—which, sadly, does not include the majority of game researchers.

Chapter 8

1. Menethil Harbor, Gadgetzan, Winterspring, and Azshara are areas in *World of Warcraft* (Blizzard Entertainment, 2004).

Appendix

1. *Instances* are areas created specifically for the group which no other players outside the group can enter. In the case of battlegrounds, players sign up by talking to the appropriate non-player character (NPC; see note 3). When enough characters from each of the two factions, Horde and Alliance, have signed up, characters get teleported to the instance and the game starts. When the game ends, they get tele-ported to a designated area in the world. No other characters can wander in while a game is going on.

2. *Mobs* are computer-controlled agents that populate MMOG landscapes, and are often intended to be killed by players in exchange for experience points.

3. *Non-player character* is a term derived from tabletop RPGs that referred to characters in the game played by the Game Master. In the context of MMOGs, NPCs are computer-controlled agents that have some degree of narrative character fleshed out.

References

Publications Cited

Aarseth, Espen. 1997. *Cybertext: Perspectives on Ergodic Literature*. Baltimore: Johns Hopkins University Press.

Aarseth, Espen. 2004. Genre Trouble. In *First Person: New Media as Story, Performance, and Game*, ed. N. Wardrip-Fruin and P. Harrigan. Cambridge, MA: MIT Press.

Aarseth, Espen. 2005. From Hunt the Wumpus to Everquest: Introduction to Quest Theory. Paper read at the International Conference on Entertainment Computing 2005, Sanda, Japan.

Babic, Edvin. 2007. On the Liberation of Space in Computer Games. *Eludamos: Journal for Computer Game Culture* 1 (1). http://www.eludamos.org/index.php/eludamos/article/viewPDFInterstitial/4/5 (accessed February 6, 2010).

Baddeley, Alan, and Graham Hitch. 1994. Development in the Concept of Working Memory. *Neuropsychology* 8 (4):485–493.

Bal, Mieke. 1997. *Narratology: Introduction to the Theory of Narrative*. Toronto: University of Toronto Press.

Bay, Michael. 2001. *Pearl Harbor*. Jerry Bruckheimer Films & Touchstone Pictures. Film.

Bazin, André. 1967. *What Is Cinema? Essays*. Berkeley: University of California Press.

Bell, John, and Zhisheng Huang. 1997. Dynamic Goal Hierarchies. In *Intelligent Agent Systems: Theoretical and Practical Issues*. Berlin: Springer.

Björk, Staffan, and Jussi Holopainen. 2005. *Patterns in Game Design*. Hingham, MA: Charles River Media.

Bolter, J. David, and Richard Grusin. 1999. *Remediation: Understanding New Media*. Cambridge, MA: MIT Press.

Borges, Jorge Luis. 1972. *Labyrinths: Selected Stories and Other Writings*. Harmondsworth: Penguin.

Brown, Emily, and Paul Cairns. 2004. A Grounded Investigation of Immersion in Games. Paper read at CHI [Conference on Human Factors in Computing Systems] 2004, Vienna.

Bryant, Jennings, and John Davies. 2006. Selective Exposure to Video Games. In *Playing Video Games: Motives, Responses, and Consequences*, ed. P. Vorderer and J. Bryant. Mahwah, NJ: Lawrence Erlbaum Associates.

Caillois, Roger. 1962. *Man, Play and Games*. London: Thames and Hudson.

Cairns, Paul, Anna Cox, Nadia Berthouze, Samira Dhoparee, and Charlene Jennett. 2006. Quantifying the Experience of Immersion in Games. Paper read at Cognitive Science of Games and Gameplay workshop, Vancouver.

Calo, M. Ryan. 2010. People Can Be So Fake: A New Dimension to Privacy and Technology Scholarship. *Penn State Law Review* 114 (3):2–50.

Carr, Diane. 2006. Play and Pleasure in Computer Games. In *Text, Narrative and Play*, ed. D. Carr, A. Burn, D. Buckingham, and G. Schott. Cambridge: Polity Press.

Castronova, Edward. 2005. *Synthetic Worlds: The Business and Culture of Online Games*. Chicago: University of Chicago Press.

Chatman, Seymour. 1978. *Story and Discourse: Narrative Structure in Fiction and Film*. Ithaca: Cornell University Press.

Copier, Marinka. 2007. Beyond the Magic Circle: A Network Perspective on Role-Play in Online Games. PhD diss., Utrecht University.

Crawford, Chris. 2003. *Chris Crawford on Game Design*. Indianapolis: New Riders.

Culler, Jonathan. 1981. *The Pursuit of Signs: Semiotics, Literature, Deconstruction*. Ithaca: Cornell University Press.

Damasio, Antonio. 2000. *The Feeling of What Happens: Body, Emotion and the Making of Consciousness*. New ed. London: Vintage.

Deleuze, Gilles, and Félix Guattari. 1987. *A Thousand Plateaus: Capitalism and Schizophrenia*. Minneapolis: University of Minnesota Press.

Dennett, Daniel C. 1991. *Consciousness Explained*. Boston: Little, Brown.

Dibbell, Julian. 2006. *Play Money: Or, How I Quit My Day Job and Made Millions Trading Virtual Loot*. New York: Basic.

Douglas, J. Yellowlees, and Andrew Hargadon. 2001. The Pleasures of Immersion and Engagement: Schemas, Scripts and the Fifth Business. *Digital Creativity* 12 (3):153–166.

Dovey, Jon, and Helen W. Kennedy. 2006. *Game Cultures: Computer Games as New Media*. Berkshire: Open University Press.

Egenfeldt-Nielsen, Simon, Jonas Heide Smith, and Susana Pajares Tosca. 2008. *Understanding Video Games: The Essential Introduction*. London: Routledge.

Ehrmann, Jacques. 1968. Homo Ludens Revisited. *Yale French Studies* 41:31–57.

Ermi, Laura, and Frans Mayra. 2005. Fundamental Components of the Gameplay Experience: Analyzing Immersion. Paper read at DIGRA 2005: Changing Views: Worlds in Play, Vancouver.

Eskelinen, Marrku. 2004. Towards Computer Game Studies. In *First Person: New Media as Story, Performance, and Game*, ed. N. Wardrip-Fruin and P. Harrigan. Cambridge, MA: MIT Press.

Evans, Andrew. 2001. *This Virtual Life: Escapism in the Media*. London: Vision.

Fan, Jin, Bruce D. McCandliss, Tobias Sommer, Amir Raz, and I. Michael Posner. 2002. Testing the Efficiency and Independence of Attentional Networks. *Journal of Cognitive Neuroscience* 14 (3):340–347.

Fernandez-Vara, Clara. 2007. Labyrinth and Maze: Videogame Navigation Challenges. In *Space Time Play: Computer Games, Architecture and Urbanism: The Next Level*, ed. F. v. Borries, S. P. Walz, and M. Bottger. Basel and Boston: Birkhäuser.

Fine, Gary Alan. 1983. *Shared Fantasy: Role-Playing Games as Social Worlds*. Chicago: University of Chicago Press.

Genette, Gérard. 1980. *Narrative Discourse: An Essay in Method*. Ithaca: Cornell University Press.

Genette, Gérard. 1988. *Narrative Discourse Revisited*. Ithaca: Cornell University Press.

Gerhard, Michael, David Moore, and Dave Hobbs. 2004. Embodiment and Copresence in Collaborative Interfaces. *International Journal of Human-Computer Studies* 61 (4):453–480.

Giddens, Anthony. 1984. *The Constitution of Society: Outline of the Theory of Structuration*. Cambridge: Polity Press.

Golledge, Reginald, and Tommy Garling. 2004. Cognitive Maps and Urban Travel. In *Handbook of Transport Geography*, ed. D. Hensher, K. Button, K. Haynes, and P. Stopher. Oxford: Elsevier.

Grau, Oliver. 2003. *Virtual Art: From Illusion to Immersion*. Cambridge, MA: MIT Press.

Grodal, Torben. 2000. Video Games and the Pleasures of Control. In *Media Entertainment: The Psychology of Its Appeal*, ed. D. Zillmann and P. Vorderer. Mahwah, NJ: Lawrence Erlbaum Associates.

Gysbers, Andre, Christoph Klimmt, Tilo Hartmann, Andreas Nosper, and Peter Vorderer. 2004. Exploring the Book Problem: Text, Design, Mental Representations of Space and Spatial Presence. Paper read at the 7th Annual International Workshop on Presence, Valencia.

Heide Smith, Jonas. 2007. Tragedies of the Ludic Commons: Understanding Cooperation in Multiplayer Games. *Game Studies* 7 (1).

Heidegger, Martin. 1993 [1977]. *The Question Concerning Technology and Other Essays*. New York: Harper & Row.

Held, Richard, and Nathaniel Durlach. 1992. Telepresence. *Presence* (Cambridge, MA) 1 (1):109–112.

Houser-Marko, Linda. 2007. Keeping Your Eyes on the Prize Versus Your Nose to the Grindstone: The Effects of Level of Goal Evaluation on Mood and Motivation. PhD diss., University of Missouri–Columbia.

Huizinga, Johan. 1955. *Homo Ludens: A Study of the Play-Element in Culture*. Boston: Beacon Press.

Hunter, Dan. 2006. An Open Letter to Blizzard: Speech Policy for GLBT guilds in World of Warcraft [February 8]. Available from http://terranova.blogs.com/terra_nova/2006/02/open_letter_to_.html.

Ijsselsteijn, Wijnand. 2004. Presence in Depth. PhD diss., Eindhoven University of Technology.

Ijsselsteijn, Wijnand, and Giuseppe Riva. 2003. Being There: The Experience of Presence in Mediated Environments. In *Being There: Concepts, Effects and Measurements of User Presence in Synthetic Environments*, ed. W. Ijsselsteijn and G. Riva. Amsterdam: Ios Press.

Iser, Wolfgang. 1991. *The Act of Reading: A Theory of Aesthetic Response*. Baltimore: Johns Hopkins University Press.

Jackson, Peter. 2001. *The Lord of the Rings: The Fellowship of the Ring*. New Zealand: New Line Cinema. Film.

Järvinen, Aki. 2008. Games without Frontiers: Theories and Methods for Game Studies and Design. PhD diss., University of Tampere, Finland.

Jenkins, Henry. 2004. Game Design as Narrative Architecture. In *First Person: New Media as Story, Performance, and Game*, ed. N. Wardrip-Fruin and P. Harrigan. Cambridge, MA: MIT Press.

Jennett, Charlene, Anna Cox, Paul Cairns, Samira Dhoparee, Andrew Epps, Tim Tijs, and Alison Walton. 2008. Measuring and Defining the Experience of Immersion in Games. *International Journal of Human-Computer Studies* 66 (9):641–661.

Juul, Jesper. 2001. Games Telling Stories? A Brief Note on Games and Narratives. *Game Studies* 1 (1).

Juul, Jesper. 2005. *Half-Real: Video Games between Real Rules and Fictional Worlds.* Cambridge, MA: MIT Press.

Kelly, Richard. 2004. *Massively Multiplayer Online Role-Playing Games: The People, the Addiction, and the Playing Experience.* Jefferson, NC: McFarland.

King, Geoff, and Tanya Krzywinska. 2006. *Tomb Raiders and Space Invaders: Videogame Forms and Contexts.* London: I. B. Tauris.

Klastrup, Lisbeth. 2004. Towards a Poetics of Virtual Worlds: Multi-User Textuality and the Emergence of Story. PhD diss., IT University of Copenhagen.

Klevjer, Rune. 2002. In Defense of Cutscenes. Paper read at Computer Games and Digital Cultures Conference, Tampere, Finland.

Koster, Raph. 2005. *A Theory of Fun for Game Design.* Scottsdale, AZ: Paraglyph Press.

Lakoff, George, and Mark Johnson. 2003. *Metaphors We Live By.* Chicago: University of Chicago Press.

Laurel, Brenda. 1991. *Computers as Theatre.* Reading: Addison-Wesley.

Lee, Kwan Min. 2004. Presence Explicated. *Communication Theory* 14 (1):27–50.

Lefebvre, Henri. 1991. *The Production of Space.* Oxford: Basil Blackwell.

Levine, Ken. 2008. Making Them Care about Your Stupid Story. Paper read at Game Developers Conference 2008, Los Angeles.

Lévy, Pierre. 1998. *Becoming Virtual: Reality in the Digital Age.* New York: Plenum.

Lewis-Williams, David. 2002. *The Mind in the Cave: Consciousness and the Origins of Art.* London: Thames & Hudson.

Liboriussen, Bjarke. 2009. The Mechanics of Place: Landscape and Architecture in Virtual Worlds. PhD diss., University of Southern Denmark.

Lim, Sohye, and Byron Reeves. 2006. Responses to Interactive Game Characters Controlled by a Computer Versus other Players. Paper read at the Annual Meeting of the International Communication Association, San Francisco.

Livingstone, Ian, and Steve Jackson. 1980. *The Warlock of Firetop Mountain.* London: Puffin Books.

Locke, Edwin A., and Gary P. Latham. 1990. *A Theory of Goal Setting and Task Performance*. Englewood Cliffs, NJ: Prentice Hall.

Lombard, Matthew, and Theresa Ditton. 1997. At the Heart of It All: The Concept of Presence. *Journal of Computer-Mediated Communication* 3 (2). http://jcmc.indiana.edu/vol3/issue2/lombard.html (accessed October 15, 2008).

Löw, Martina. 2001. *Raumsoziologie. 1*. Frankfurt: Suhrkamp.

Lynch, Kevin. 1960. *The Image of the City*. Cambridge, MA: MIT Press and Harvard University Press.

Mackay, Daniel. 2001. *The Fantasy Role-Playing Game: A New Performing Art*. Jefferson, NC: McFarland.

The Making of *Grand Theft Auto IV*. 2008. *Edge*, April.

Malaby, Thomas. 2007. Beyond Play: A New Approach to Games. *Games and Culture* 2 (2):95–113.

Manninen, Tony, and Tomi Kujanpaa. 2005. The Hunt for Collaborative War Gaming—CASE: *Battlefield 1942*. *Game Studies* 5 (1).

Marsh, Tim. 2003. Presence as Experience: Film Informing Ways of Staying There. *Presence* (Cambridge, MA) 12 (5):538–540.

Mateas, Michael, and Andrew Stern. 2005. Build It to Understand It: Ludology Meets Narratology in Game Design Space. Paper read at DIGRA [Digital Games Research Association] 2005: Changing Views: Worlds in Play, Vancouver.

McIntosh, William. D. 1996. When Does Goal Non-Attainment Lead to Negative Emotional Reactions and When Doesn't It? The Role of Linking and Rumination. In *Striving and Feeling: Interactions among Goals, Affect and Self-Regulation*, ed. L. L. Martin and A. Tesser. Mahwah, NJ: Lawrence Erlbaum Associates.

McMahan, Alison. 2003. Immersion, Engagement and Presence: A Method for Analyzing 3-D Video Games. In *The Video Game Theory Reader*, ed. M. J. P. Wolf and B. Perron. London: Routledge.

Metz, Christian. 1974. *Film Language: A Semiotics of the Cinema*. New York: Oxford University Press.

Minsky, Marvin. 1980. Telepresence. *Omni Magazine*.

Morris, Sue. 2002. First Person Shooters: A Game Apparatus. In *Screenplay: Cinema/Videogames/Interfaces*, ed. G. King and T. Krzywinska. London: Wallflower.

Murray, Janet Horowitz. 1998. *Hamlet on the Holodeck: The Future of Narrative in Cyberspace*. Cambridge, MA: MIT Press.

Næss, Arne. 2005. *The Selected Works of Arne Næss*. Ed. Alan Drengson. Dordrecht: Springer.

Nell, V. 1988. *Lost in a Book: The Psychology of Reading for Pleasure*. New Haven: Yale University Press.

Nitsche, Michael. 2008. *Video Game Spaces: Image, Play, and Structure in 3D Game Worlds*. Cambridge, MA: MIT Press.

Oldenburg, Ray. 1999. *The Great Good Place: Cafés, Coffee Shops, Bookstores, Bars, Hair Salons, and Other Hangouts at the Heart of a Community*. Berkeley, CA: Marlowe.

Oxford Dictionary of English. 2003. 2nd ed. Ed. Catherine Soanes and Angus Stevenson. Oxford: Oxford University Press.

Pargman, Daniel, and Peter Jakobsson. 2006. The Magic Is Gone: A Critical Analysis of the Gaming Situation. Paper read at Medi@terra—Gaming Realities: A Challenge for Digital Culture, Athens, Greece.

Pearce, Celia. 2004. Towards a Game Theory of Game. In *First Person: New Media as Story, Performance, and Game*, ed. N. Wardrip-Fruin and P. Harrigan. Cambridge, MA: MIT Press.

Poels, Karolien. 2009. *World of Warcraft*, the Aftermath: How Game Elements Transfer into Real Life Perceptions and Experiences. Paper read at DIGRA [Digital Games Research Association] 2009: Breaking New Ground—Innovations in Games, Play, Practice, and Theory, Brunel, UK.

Poels, Karolien, Wijnand Ijsselsteijn, Yvonne De Kort, and B. Van Iersel. 2010. Digital Games, the Aftermath: Qualitative Insights into Post Game Experiences. In Evaluating User Experiences in Games, ed. R. Bernhaupt. Berlin: Springer.

Prince, Gerald. 1982. *Narratology: The Form and Functioning of Narrative*. Janua linguarum. Series maior, 108. Berlin: Mouton.

Proust, Marcel. 1954. *À la recherche du temps perdu*. 8 vols. Paris: Gallimard.

Rambusch, Jana, Daniel Pargman, and Peter Jakobsson. 2007. Exploring E-sports: A Case Study of Gameplay in *Counter-Strike*. Paper read at DIGRA [Digital Games Research Association] 2007: Situated Play, Tokyo, Japan.

Reeves, Byron, and Clifford Nass. 1996. *The Media Equation: How People Treat Computers, Television, and New Media Like Real People and Places*. Cambridge: Cambridge University Press.

Rettie, Ruth. 2004. Using Goffman's Frameworks to Explain Presence and Reality. Paper read at the 7th Annual International Workshop on Presence, Valencia, Spain.

Richards, Ivan A. 1936. *The Philosophy of Rhetoric*. New York: Oxford University Press.

Rollings, Andrew, and Dave Morris. 2000. *Game Architecture and Design*. Scottsdale, AZ: Coriolis.

Rosenbloom, Paul, and Alen Newell. 1983. The Chunking of Goal Hierarchies: A General Model of Practice. In *Machine Learning: An Artificial Intelligence Approach*, ed. J. R. Anderson, R. S. A. Michalski, J. G. Carbonell, T. M. Mitchell, Y. Kodratoff, and G. Tecuci. Los Altos, CA: M. Kaufmann.

Ruvinsky, Alicia, and Michael N. Huhns. 2009. Human Behavior in Mixed Human-Agent Societies. In *Proceedings of the 8th International Conference on Autonomous Agents and Multiagent Systems*, vol. 2. Budapest: International Foundation for Autonomous Agents and Multiagent Systems.

Ryan, Marie-Laure. 2001. *Narrative as Virtual Reality: Immersion and Interactivity in Literature and Electronic Media*. Baltimore: Johns Hopkins University Press.

Ryan, Marie-Laure. 2006. *Avatars of Story*. Electronic Mediations, 17. Minneapolis: University of Minnesota Press.

Salen, Katie, and Eric Zimmerman. 2003. *Rules of Play: Game Design Fundamentals*. Cambridge, MA: MIT Press.

Sartre, Jean-Paul. 1967. *What Is Literature?* London: Methuen.

Sartre, Jean-Paul. 1995 [1948]. *The Psychology of Imagination*. London: Routledge.

Schank, Roger C., and Robert P. Abelson. 1977. *Scripts, Plans, Goals, and Understanding: An Inquiry into Human Knowledge Structures*. Hillsdale, NJ: LEA.

Schott, Gareth. 2006. Agency in and around Play. In *Computer Games: Text, Narrative, and Play*, ed. D. Carr, A. Burn, D. Buckingham, and G. Schott. Cambridge: Polity Press.

Schubert, Thomas, and Jan Crusius. 2002. Five Theses on the Book Problem. Available from http://www.igroup.org/projects/porto2002/SchubertCrusiusPorto2002 .pdf (accessed July 15, 2006).

Sheridan, Thomas B. 1992. Musings on Telepresence and Virtual Presence. *Presence* (Cambridge, MA) 1 (1).

Slater, Mel. 2003. A Note on Presence Terminology. Available from http://presence .cs.ucl.ac.uk/presenceconnect (accessed October 15, 2008).

Slater, Mel, and S. Wilbur. 1997. A Framework for Immersive Virtual Environments (Five): Speculations on the Role of Presence in Virtual Environments. *Presence* (Cambridge, MA) 6 (6):603–616.

Soja, Edward W. 1996. *Thirdspace: Journeys to Los Angeles and Other Real-and-Imagined Places*. Cambridge, MA: Blackwell.

Spielberg, Stephen. 1998. *Saving Private Ryan*. Dreamworks. Film.

Steinkuehler, Constance. 2005. Cognition and Learning in Massively Multiplayer Online Games: A Critical Approach. PhD diss., University of Wisconsin.

Steinkuehler, Constance, and Dmitri Williams. 2006. Where Everybody Knows Your (Screen) Name: Online Games as "Third Places." Available from http://website .education.wisc.edu/steinkuehler/papers (accessed December 10, 2006).

Steuer, Jonathan. 1992. Defining Virtual Reality: Dimensions Determining Telepresence. *Journal of Communication* 42 (4):73–93.

Sturm, W., and K. Willmes. 2001. On the Functional Neuroanatomy of Intrinsic and Phasic Alertness. *NeuroImage* 14 (1):76–84.

Suits, Bernard Herbert. 1978. *The Grasshopper: Games, Life, and Utopia.* Toronto: University of Toronto Press.

Sutherland, Ivan. 1965. The Ultimate Display. Paper read at IFIPS [International Federation of Information Processing Societies] congress, New York.

Sutherland, Ivan. 1968. Head-Mounted Three-Dimensional Display. Paper read at Fall Joint Computer Conference, San Francisco.

Swalwell, Melanie. 2008. Movement and Kinesthetic Responsiveness: A Neglected Pleasure. In *The Pleasures of Computer Gaming: Essays on Cultural History, Theory, and Aesthetics*, ed. M. Swalwell and J. Wilson. Jefferson, NC: McFarland.

Tamborini, Ron, and Paul Skalski. 2006. The Role of Presence in the Experience of Electronic Games. In *Playing Video Games: Motives, Responses, and Consequences*, ed. P. Vorderer and J. Bryant. Mahwah, NJ: Lawrence Erlbaum Associates.

Taylor, T. L. 2006. *Play between Worlds: Exploring Online Game Culture.* Cambridge, MA: MIT Press.

Tolkien, J. R. R. 1983. *The Monsters and the Critics and Other Essays.* Ed. Christopher Tolkien. London: Allen & Unwin.

Tolman, Edward. 1948. Cognitive Maps in Rats and Men. *Psychological Review* 55 (4):189–208.

Tuan, Yi-Fu. 1974. *Topophilia: A Study of Environmental Perception, Attitudes, and Values.* Englewood Cliffs, NJ: Prentice-Hall.

Tuan, Yi-Fu. 1998. *Escapism.* Baltimore: Johns Hopkins University Press.

Tuan, Yi-Fu. 2007. *Space and Place: The Perspectives of Experience.* Minneapolis: University of Minnesota Press.

Van den Hoogen, Wouter, Wijnand Ijsselsteijn, and Yvonne de Kort. 2009. Effects of Sensory Immersion on Behavioral Indicators of Player Experience: Movement Synchrony and Controller Pressure. Paper read at DIGRA [Digital Games Research

Association] 2009: Breaking New Ground—Innovations in Games: Play, Practice and Theory, Brunel, UK.

Van Rooyen-McLeay, Karen. 1985. Adolescent Video-Game Playing. Master's thesis, Victoria University of Wellington.

Wachowski, Larry, and Andy Wachowski. 1999. *The Matrix*. Warner Brothers Pictures. Film.

Walsh, Richard. 2007. *The Rhetoric of Fictionality: Narrative Theory and the Idea of Fiction*. Columbus: Ohio State University Press.

Walton, Kendall L. 1990. *Mimesis as Make-Believe: On the Foundations of the Representational Arts*. Cambridge, MA: Harvard University Press.

Waterworth, John A., and Eva L. Waterworth. 2003. The Core of Presence: Presence as Perceptual Illusion. *Presence Connect*, http://presence.cs.ucl.ac.uk/presenceconnect/index.html (accessed November 12, 2008).

Witmer, B. G., and M. J. Singer. 1998. Measuring Presence in Virtual Environments: A Presence Questionnaire. *Presence* (Cambridge, MA) 7 (3):225–240.

Wittgenstein, Ludwig. 1997 [1953]. *Philosophical Investigations*. Ed. G. Anscombe. 2d ed. Cambridge, MA: Blackwell.

Wolf, Mark J. P. 2009. Theorizing Navigable Spaces in Video Game Worlds in Structure, Sign, and Play. Paper presented at DIGAREC [Digital Games Research Center], University of Potsdam, Germany.

World of Warcraft: Terms of Use 2006 [cited February 1]. Blizzard Entertainment. Available at http://www.worldofwarcraft.com/legal/termsofuse.html.

Yee, Nick. 2006a. The Demographics, Motivations, and Derived Experiences of Users of Massively-Multiuser Online Graphical Environments. *Presence* (Cambridge, MA) 15:309–329.

Yee, Nick. 2006b. The Labor of Fun. *Games and Culture* 1 (1):68–71.

Games Discussed

Age of Conan. Funcom, 2006 [cited December 15]. Available from www.ageofconan.com.

Age of Empires. Ensemble Studios (Microsoft Game Studios), 1997. PC.

Aion: The Tower of Eternity. NC Soft, 2009. PC.

Asheron's Call. Turbine Inc. (Microsoft), 1999. PC.

Asheron's Call 2. Turbine Inc., 2002. PC.

Assassin's Creed. Ubisoft Montreal, 2007. Xbox 360.

Battlefield 1942. Digital Illusions CE, 2002. PC.

Bejewelled. PopCap Games, 2001. PC.

Bioshock. 2K Games, 2008. Xbox 360.

Call of Duty: Modern Warfare 2. Infinity Ward, 2009. PC.

Call of Duty IV. Infinity Ward, 2007. PC.

Combat. Atari Inc., 1977. Atari 2600.

Counter-Strike. Valve Software (Sierra Studios), 1998. PC.

Counter-Strike: Source. Valve Software (Vivendi Universal Games), 2004. PC.

Crysis. Crytek Frankfurt, 2007. PC.

Dark and Light. NP Cube (Farlan), 2006. PC.

Dead Space. EA Redwood Shores (Electronic Arts), 2008. Xbox 360.

Dear Esther. The Chinese Room, 2008. www.moddb.com. PC.

Doom. id Software (id Software & GT Interactive), 1993. PC.

Dungeons and Dragons. David Arneson and Gary Gygax (Tactical Studies Rules), 1974.

The Elder Scrolls IV: Oblivion. Bethesda Softworks LLC (2K Games), 2006. PC.

Eliss. Steph Thirion (Apple), 2009. iPhone.

Empire: Total War. Creative Assembly (Sega), 2009. PC.

Enter the Matrix. Shiny Entertainment (Atari & Warner Brothers Interactive), 2003. PC.

EVE Online. CCP Games, 2003. PC.

EverQuest. Sony Online Entertainment, 1999. PC.

Fable. Lionhead Studios (Microsoft Game Studios), 2004. Xbox 360.

Fable II. Lionhead Studios (Microsoft Game Studios), 2008. Xbox 360.

Face of Mankind. Duplex Systems (Ojom), 2006. PC.

Fahrenheit. Quantic Dream (Atari), 2005. PC.

Fallout 3. Bethesda Game Studios, 2008. PC.

Far Cry 2. Ubisoft Montreal, 2008. PC.

F.E.A.R. Monolith Inc. (Vivendi Universal), 2005. PC.

F.E.A.R. 2: Project Origin. Monolith, Inc. (Warner Brothers Interactive Entertainment), 2009. PC.

FIFA 2008. EA Sports, 2007. XBOX 360.

FIFA 2009. EA Sports, 2008. XBOX 360.

Flower. ThatGameCompany (Sony Computer Entertainment), 2009. PS3.

Grand Slam Tennis. EA Canada (EA Sports), 2009. Wii.

Grand Theft Auto IV. Rockstar North (Rockstar Games), 2008. Xbox 360.

Gran Turismo. Polyphony Digital (Sony Computer Entertainment), 1998. PS.

Guild Wars. ArenaNet (NC Soft), 2005. PC.

Half-Life 2. Valve Software (Vivendi Universal), 2004. PC.

Halo 3. Bungie (Microsoft Game Studios), 2007. Xbox 360.

Home. SCE London Studio (Sony Computer Entertainment), 2008. PS3.

The House of the Dead. Wow Entertainment (Sega), 1996. PC.

Huxley. Webzen Inc., forthcoming. PC.

Kick-Off. Dino Dini (Anco), 1989. Amiga.

Left 4 Dead. Valve Corporation, 2008. PC.

Left 4 Dead 2. Valve Corporation, 2009. PC.

Lineage. NC Soft, 1998. PC.

Lineage 2. NC Soft, 2003. PC.

The Lord of the Rings Online: Shadows of Angmar. Turbine Inc., 2007. PC.

Mass Effect. Bioware (Microsoft Game Studios), 2007. Xbox 360.

Max Payne. Rockstar Toronto (Rockstar Games), 2001. PS2.

Medal of Honor: Allied Assault. 2015, Inc. (Electronic Arts), 2002. PC.

Medal of Honor: Pacific Assault. Electronic Arts and TKO Software (Electronic Arts), 2004. PC.

Medieval II: Total War. Creative Assembly (Sega), 2006. PC.

Metal Gear Solid 4. Kojima Productions (Konami), 2008. PS3.

Mirror's Edge. DICE, 2008. PC.

Mount and Blade. Tale Worlds (Paradox Interactive), 2008. PC.

Myst III. Presto Studios (Ubisoft), 2001. PC.

Need for Speed Underground 2. Electronic Arts, 2006. Available from http://www.ea.com/nfs/underground2/us/home.jsp (accessed December 15, 2006).

No More Heroes. Grasshopper Manufacture (Rising Star Games), 2008. Wii.

Pacman. Namco, 1980. Arcade.

Planetside. Sony Online Entertainment, 2003. PC.

Pong. Atari Inc., 1972. Arcade.

Red Orchestra: Ostfront 41–45. Tripwire Interactive (Valve Software), 2006. PC.

Risk: Revised Edition. Hasbro, 2009. Board game.

Rock Band. Harmonix (MTV Games), 2007. Xbox 360.

Second Life. Linden Lab, 2003. PC.

The Secret of Monkey Island. Lucasfilm Games, 1990. Amiga.

Settlers of Catan. Claus Teuber (Mayfair Games), 1995.

SimCity. Maxis Software (Electronic Arts), 1989. PC.

The Sims. Maxis Software (Electronic Arts), 2000. PC.

Space Invaders. Taito (Midway), 1978. Arcade.

Spacewar. Steve Russell, 1962. Mainframe.

Spore. Maxis Software (Electronic Arts), 2008. PC.

S.T.A.L.K.E.R.: Shadow of Chernobyl. GSC Gameworld (THQ), 2007. PC.

Tetris. A. Pajitnov (Spectrum Holobyte), 1985. PC.

Unreal Tournament. Epic Games (GT Interactive), 1999. PC.

Warhammer 40,000: Dawn of War. Relic Entertainment (THQ), 2004. PC.

Warhammer 40,000: Dawn of War II. Relic Entertainment (THQ), 2009. PC.

Wii Sports. Nintendo EAD, 2006. Wii.

Wolfpack. Simulations Publications Inc., 1975. Board game.

World of Warcraft. Blizzard Entertainment (Vivendi Universal), 2004. PC.

World of Warcraft Map. 2006. Available from http://mapwow.com (accessed February 10, 2006).

World War II Online. Corner Rat Software and Playnet Inc. (GMX Media), 2001. PC.

Index